Communication and Networking in Smart Grids

Edited by Yang Xiao

CRC Press
Taylor & Francis Group
Boca Raton   London   New York

CRC Press is an imprint of the
Taylor & Francis Group, an **informa** business

Contents

PART II COMMUNICATIONS AND NETWORKS IN SMART GRIDS

Preface

Smart grids are an integration of power delivery systems with communication networks and information technology (IT) to provide better services. Communication and networking will provide significant roles in building future smart grids. The purpose of this book is to provide state-of-the-art approaches and novel technologies for communication networks in smart grids, covering a range of topics in the areas, making it an excellent reference book for students, researchers, and engineers in these areas.

This book investigates fundamental aspects and applications of smart grids, its communications, and networks. It presents a collection of recent advances in these areas. Many prominent researchers working on smart grids and related fields around the world have contributed to this work. The book contains 12 chapters, that are divided into two parts: "Smart Grids in General" and "Communications and Networks in Smart Grids." We believe this book will be a solid reference tool for researchers, practitioners, and students who are interested in the research, development, design, and implementation of smart grid communications and networks.

This book was made possible by the great efforts of our contributors and publishers. We are indebted to our contributors, who have sacrificed their valuable time to put together these chapters for our readers. We thank our publisher Taylor & Francis—without their encouragement and quality work, this book would not be possible.

Yang Xiao
Department of Computer Science
University of Alabama
E-mail: yangxiao@ieee.org

Acknowledgment

This work is supported in part by the U.S. National Science Foundation (NSF) under grants CCF-0829827, CNS-0716211, CNS-0737325, and CNS-1059265.

About the Editor

Dr. Yang Xiao worked in industry as a medium access control (MAC) architect involving IEEE 802.11 standard enhancement work before he joined the Department of Computer Science at the University of Memphis in 2002. He is currently a tenured professor in the Department of Computer Science at the University of Alabama. He was a voting member of IEEE 802.11 working group from 2001 to 2004, and is currently an IEEE senior member. He serves as a panelist for the U.S. National Science Foundation (NSF), Canada Foundation for Innovation's (CFI) Telecommunications expert committee, and the American Institute of Biological Sciences (AIBS), as well as a referee/reviewer for many national and international funding agencies. His areas of research are security, communications/networks, robotics, and telemedicine. He has published more than 180 refereed journal papers and over 200 refereed conference papers and book chapters related to these research areas. Dr. Xiao's research has been supported by the U.S. National Science Foundation (NSF), U.S. Army Research, Global Environment for Network Innovations (GENI), Fleet Industrial Supply Center–San Diego (FISCSD), FIATECH, and the University of Alabama's Research Grants Committee. He currently serves as editor-in-chief for the *International Journal of Security and Networks (IJSN)* and the *International Journal of Sensor Networks (IJSNet)*. He was the founding editor-in-chief for the *International Journal of Telemedicine and Applications (IJTA)* (2007–2009).

Contributors

M. Cheriet
Ecole de Technogie Superieure
University of Quebec
Montreal, Quebec, Canada

Philippe Daniel
Accenture Technology
Sophia Antipolis, France

A. Daouadji
Ecole de Technogie Superieure
University of Quebec
Montreal, Quebec, Canada

Debraj De
Sensorweb Research Laboratory
Department of Computer Science
Georgia State University
Atlanta, Georgia

Tomaso Erseghe
Department of Information
    Engineering
University of Padova
Padova, Italy

Lorenza Giupponi
Centre Tecnológic de
    Telecomunicacions de
    Catalunya (CTTC)
Barcelona, Spain

Juan José Gonzalez de la Rosa
Universidad de Cadiz
Andalusia, Spain

M. Gonzalez-Redondo
Department of Computer
    Architecture, Electronics,
    and Electronic Technology
University of Cardoba
Cordoba, Spain

David Gregoratti
Centre Tecnológic de
    Telecomunicacions de
    Catalunya (CTTC)
Barcelona, Spain

Dong Han
Intelligent Sensor Grid and
    Informatics Lab
University of Houston
Houston, Texas

Christian Ibars
Centre Tecnológic de
    Telecomunicacions de
    Catalunya (CTTC)
Barcelona, Spain

Scott Kurth
Accenture Technology Vision
Chicago, Illinois

M. Lemay
Inocybe Technologies, Inc.
Montreal, Canada

Gang Lu
Sensorweb Research Laboratory
Department of Computer Science
Georgia State University
Atlanta, Georgia

Javier Matamoros
Centre Technologic de
    Telecommunications
    de Catalunya (CTTC)
Barcelona, Spain

I.M. Moreno-Garcia
Department of Computer
    Architecture, Electronics,
    and Electronic Technology
University of Cordoba
Cordoba, Spain

A. Moreno-Munoz
Department of Computer
    Architecture, Electronics,
    and Electronic Technology
University of Cordoba
Cordoba, Spain

Christian Müller
TU Dortmund University
Communication Networks Institute
Dortmund, Germany

Monica Navarro
Centre Tecnológic de
    Telecomunicacions de
    Catalunya (CTTC)
Barcelona, Spain

K.-K. Nguyen
Ecole de Technogie Superieure
University of Quebec
Montreal, Quebec, Canada

Víctor Pallarés-López
Department of Computer
    Architecture, Electronics,
    and Electronic Technology
University of Cordoba
Cordoba, Spain

Matthias Postina
R&D Division Energy
OFFIS
Oldenburg, Germany

R. Real-Calvo
Department of Computer
    Architecture, Electronics,
    and Electronic Technology
University of Cordoba
Cordoba, Spain

Sebastian Rohjans
R&D Division Energy
OFFIS
Oldenburg, Germany

Jens Schmutzler
TU Dortmund University
Communication Networks Institute
Dortmund, Germany

Aline Senart
Accenture Technology Labs
Sophia Antipolis, France

Alberto Sendin
Faculty of Engineering
Department of Telecommunications
University of Deusto
Bilbao, Spain

Autumn Nicole Smith
University of Alabama
Tuscaloosa, Alabama

Wen-Zhan Song
Sensorweb Research Laboratory
Department of Computer Science
Georgia State University
Atlanta, Georgia

Christian Souche
Accenture Technology Labs
Sophia Antipolis, France

Michael Specht
R&D Division Energy
OFFIS
Oldenburg, Germany

Ulrike Steffens
R&D Division Energy
OFFIS
Oldenburg, Germany

Wei Sun
Department of Electrical
    Engineering
Hefei University of Technology
Hefei, China

Paolo Tenti
Department of Information
    Engineering
University of Padova
Padova, Italy

Stefano Tomasin
Department of Information
    Engineering
University of Padova
Padova, Italy

Joern Trefke
R&D Division Energy
OFFIS
Oldenburg, Germany

Mathias Uslar
R&D Division Energy
OFFIS
Oldenburg, Germany

Jianping Wang
Department of Electrical Engineering
Hefei University of Technology
Hefei, China

Christian Wietfeld
TU Dortmund University
Communication Networks Institute
Dortmund, Germany

Yang Xiao
Department of Computer Science
University of Alabama
Tuscaloosa, Alabama

Susumu Yoneda
Softbank Telecom Corp.
Tokyo, Japan

Xiaohui Yuan
Computer Science and Engineering
    Department
University of North Texas
Denton, Texas

Xiaojing Yuan
Intelligent Sensor Grid and
    Informatics Lab
University of Houston
Houston, Texas

Chongwei Zhang
Department of Electrical
    Engineering
Hefei University of Technology
Hefei, China

Chapter 1

Smart Grids

Autumn Nicole Smith and Yang Xiao

Contents

Prof. Yang Xiao is the corresponding author, e-mail: yangxiao@ieee.org.

Our current grid system is quickly becoming obsolete. This grid system will not be able to meet our future electricity demands. New efficient technology must be introduced to solve this problem. One solution to this problem is smart grid. Smart grids will be able to efficiently handle our increasing energy demands and reduce the environmental impact by incorporating renewable resources. In this chapter, we will discuss why smart grids are vital to our future, the different types of new technology that they are comprised of, the current advancements, and research that is being conducted.

1.1 Introduction

The North American power grid has made few advances in the past century. The current grid is unable to meet the growing demand for energy. Grid congestion and congested transmission lines are becoming more frequent across the country. These issues can be addressed by implementing smart grid technology. Smart grids will be able to monitor and control the flow of electricity in real time. Smart grids will apply our new developments in information management and automation technology to the existing grid. They will also offer more control and be able to process more information, which will provide many benefits to consumers. These smart grids would provide a more efficient, reliable, environmentally friendly, and secure alternative to our current grid system.

1.1.1 Efficiency and Reliability

Our current grid system is unable to efficiently supply the energy needed by our country. Heavily populated areas in the United States are often plagued with blackouts and congested transmission lines, also known as *bottlenecks*. The U.S. Department of Energy reports that these power disturbances and outages cost the country "from $25 to 180 million" every year [1]. With an ever-increasing population and advancements in technology, there is a greater demand for a more resourceful and reliable power grid. The energy consumption rate has risen from 10% in 1940 to 40% in 2003 [1]. These increasing failures of our current grid come at a time when the demand for electricity is the highest. In the past decade, our society has been increasingly digitalized. We are more dependent on electricity than ever before. There is a rapidly growing market of technologies that rely on electricity. Just as communication has evolved with technologies such as the Internet and wireless cell phones, our power grid must also evolve. Our current grid will have to evolve quickly to combat the changes that are taking place in our increasingly digital society. Smart grids will be able to monitor energy usage in real time and predict outages and equipment failures.

A smart grid will be able to restore itself after a blackout or a weather-related outage.

1.1.2 Environmental Benefits

Our aging grid relies heavily on dwindling fossil energy resources. The cost of natural gas, coal, and oil is rising steadily. Oil, specifically, has seen a sharp price increase. Prices for oil have increased 800% from 1998 to 2008 [2]. Using these fossil fuels to supply energy to our country also contributes to our growing climate change problem. A cleaner alternative to fossil fuels is needed, and one can be implemented by using our readily available renewable resources. These resources reduce the amount of pollution being expelled into the atmosphere. Wind and solar energy production sites are being placed in remote locations and even offshore. Transporting electricity over long distances, thus far, has proved to be inefficient. Smart grids will be able to efficiently transport energy over vast distances, and therefore utilize the energy from distant renewable resources.

Recently, President Barack Obama has addressed our current energy and environmental issues [3]. In his energy plan, he calls for $3.4 billion of federal stimulus funds to be invested to modernize our current grid system by specifically using smart grids [3]. This investment will serve as a downpayment on our future grid system. President Obama is also an advocate of smart metering and has made it a point to invest in smart meter technology for American homes.

1.1.3 Benefits to Consumers

Smart grids will also help consumers save money. Consumers will be able to monitor their home usage by using smart meters. This will encourage consumers to use less energy and will reduce the amount of overall energy needed by the grid. Using less energy at times of peak demand saves money for the consumer. This is because energy produced in periods of high demand costs more to produce than energy produced in times of low demand. Consumers will actively help balance supply and demand and increase reliability by changing the way they use and purchase electricity. Allowing consumers to see the real-time price of electricity will make them more conscious of their usage. Reducing the demand for energy needed from the grid also reduces the amount of pollution being created. Many jobs will be created during both the production (i.e., planning and construction) and postproduction (i.e., maintenance, development, etc.) stages.

Smart grids will be compatible with electric vehicles. Smart grids will be able to use electric cars as energy storage by drawing power from the charging cars when demand is peaking. Thousands of electric cars charging

these renewable resources are located in remote areas. One example of this is wind energy. Most wind energy is located away from metropolitan areas in the western area of the United States. By integrating more renewable energy sources into our grid, we not only increase the power we are able to supply but also reduce our usage of fossil fuels and our impact on the environment. Another benefit of superconducting cables is the reliability that they offer by using superconductor-based fault current limiters. Superconductor-based fault current limiters "limit the amount of current flowing through the system and allow for the continual, uninterrupted operation of the electrical system" [8]. Fault current limiters function similarly to surge protectors. The excessive current produced during a fault would quickly be controlled by the fault current limiter and would thus be stopped from causing any future damage. The fault current limiter would still allow for normal current to pass through the cables. In addition to superconducting cables, high-voltage direct-current (HVDC) cables will be used to transfer electricity over long distances. HVDC systems are more cost efficient and stable than alternating current (AC) systems when transferring power over long distances. However, AC systems are cheaper for transferring electricity over short distances. HVDC will use fault current limiters as well to protect against blackouts and the damage that they can cause. HVDC systems will work with flexible alternating current transmission system (FACTS) devices. These power flow devices will help monitor the voltage on the grid in real time. They will be able to monitor the power being transferred over the superconducting cables to the power plants [9]. They will also be able to increase power transfer levels. Using HVDC and FACTS will increase transmission distance, reduce line loss, protect against blackouts, utilize renewable energy sources, and increase stability along the grid [38].

1.2.4 Energy Storage

Energy storage is another vital component of the smart grid. Energy storage will be placed throughout the grid. This is known as distributing energy resources. This will make it easier for the grid to supply emergency power during a demand increase and to release congestion along the grid [4]. This will also diminish the use of expensive backup plants that utilities use during peak load periods [10]. Energy storage is also a fundamental component of renewable energy. Due to the intermittent nature of renewable energy resources, storage of this energy is greatly needed. The smart grid will be able to store wind and solar energy and save it for later use during peak energy periods. Using this energy storage will reduce the capacity needed by transmission lines [31]. Using energy storage will increase the reliability and efficiency of the power grid while reducing the impact that it has on the environment.

Smart grids will incorporate fuel cells into the grid system. A fuel cell is a "device capable of generating an electrical current by converting the chemical energy of a fuel directly into electrical energy" [14]. Fuel cells do not contain active materials required by the electrochemical conversion inside them. These active materials are supplied from the outside. Fuel cells also have a very small carbon footprint [18]. Fuel cells can be recharged easily by adding more solution, and they never run down like traditional batteries. Fuel cells will be able to reduce peak load problems because of their high energy density and reliability. Combined with electrolyzers, fuel cells will be able to store energy. This is done by converting electrical energy to a fuel, usually hydrogen. That fuel is then used to power fuel cells to create energy when needed. Hydrogen fuel cells are particularly useful because they are not affected by temperature, they can withstand long periods of storage, and they can be cycled many times [30]. Hydrogen fuel cells have not been used yet for energy storage because of their cost. Polymer electrolyte membrane (PEM) fuel cells may provide a cost-efficient solution in the near future. PEM fuel cells are commonly used in automobiles but are being researched as an energy storage solution for the power grid. PEM fuel cells will use hydrogen, oxygen, and water to produce electricity and create water as a by-product of the reaction. New developments in self-pressuring PEM electrolyzers and reversible PEM fuel stacks have reduced the cost of hydrogen fuel cells greatly. Fuel cells are useful for storing energy from renewable resources, such as solar and wind energy, because they do not produce consistent amounts of electricity. Hydrogen fuel cells will assist in load management as well by providing energy during peak periods. Hydrogen fuel cells are already being used as power storage on the Indonesian island of Sumatra by Hutchinson Telecommunications. Hutchinson uses these cells to provide backup for their telecom network when power from the grid is lost. These cells, developed by IdaTech, are either fueled directly by hydrogen or by using a liquid fuel consisting of methanol and water that is converted into hydrogen gas to power the system [32].

Ultracapacitors will also be used for storage. Ultracapacitors are DC energy sources that are able to recharge and discharge quickly and can handle very high currents. Unlike batteries, ultracapacitors store energy electrostatically. This process is drastically faster than the chemical process used in traditional batteries. They can be recharged many times but can only store a limited amount of power. Ultracapacitors will be able to store energy for short periods of time and reduce load leveling problems. They will work along with batteries to maintain load levels along the grid. They will be used as an uninterrupted power supply (UPS) to supply emergency power. Two major drawbacks of using a system of ultracapicators for wide-scale energy storage are their high cost and low energy storage density.

The grid system of the future will encompass many different types of batteries for energy storage. Battery storage has been notorious for its costly nature, but recently new breakthroughs have been made to reduce these costs. Many different types of batteries will be used together to provide storage for the grid. These batteries include lead acid, sodium sulfur, and lithium ion batteries.

Lead acid and sodium sulfur batteries have been the most commonly used batteries for grid storage. Lead acid batteries are becoming increasingly outdated due to their low energy density and their large size. New battery technologies, such as lead-carbon batteries, have proven to be "three to four times" denser than normal lead acid batteries [31]. Sodium sulfur batteries offer a better alternative to lead acid batteries. Sodium sulfur batteries store and create energy through the chemical reaction of sodium polysulphide and sodium. This process can be repeated thousands of times. The drawback to sodium sulfur batteries is that the reactants have to be stored separately at very high temperatures. If the batteries are allowed to cool down and discharge they become inoperable. Sodium sulfur batteries are currently being used in Nagoya, Japan for grid storage. This plant can create about 300 MW of extra power when needed for up to 6 hours [29]. General Electric plans to construct a large sodium sulfur battery factory in the United States in 2009. This factory intends to have a capacity of 900 mWh/year [31]. Lithium-ion batteries are also being researched as prospective energy storage options. Lithium-ion batteries have a very high energy density and are highly efficient. However, lithium-ion batteries have thus far proven to be costly in comparison with lead acid and sodium sulfur batteries.

In addition to these technologies, the smart grid will incorporate new ways of generating energy and storing it for later use. One such technology is compressed air. Compressed air energy storage takes off-peak energy from power plants to pump air into an underground air storage tank where it is stored under pressure. This air is then released during periods of peak demand. This released air powers a wind turbine and creates energy. One such compressed air plant in Hurntorf, Germany, is capable of producing up to 300 MW of reserve power for up to 3 hours [29]. Another energy storage technology that will be incorporated is pumped hydrogen. This technology has been around for many years and "is the most widespread energy storage system in use on power networks" [19]. Pumped hydrogen works by using two reservoirs of water, one being at a higher elevation than the other. During off-peak hours, water is pumped into the higher reservoir. When energy is needed, water is released from the top reservoir and flows downward through turbines to the lower reservoir. Pumped-storage hydroelectricity can be used to reduce peak load problems. Another way of storing energy is flywheels. Flywheels take excess energy and convert it into kinetic energy by running it through a motor that spins the flywheel. By slowing down the generator, flywheels give back the energy. Flywheels can absorb energy quickly and release it just as fast. Flywheels need very little upkeep and do not contain the toxic chemicals that some batteries do [10].

1.3 First Smart Grids/Current Attempts

Even though a countrywide smart grid is still many years away, many cities and countries around the world have begun smart grid experiments. In the United States, Austin, Texas, and Boulder, Colorado, have both started smart grid experiments. Ontario, Canada, is currently developing a new grid system with smart grid technology. Italy has been using smart meters since 2001 and continues to make improvements to its grid system. These smart grid experiments are vital to revolutionizing our grid system. By testing smart grids on a small scale, we can learn more about the problems and issues a larger smart grid will face and then use this new information to further develop a wide-scale smart grid solution.

1.3.1 Boulder, CO

The Boulder smart grid project was started by Xcel Energy in 2008. The $100 million Boulder smart grid project has been dubbed SmartGridCity [20]. SmartGridCity is a "fully integrated smart grid community with what is possibly the densest concentration of these emerging technologies to date" [20]. SmartGridCity is the first of its kind in the world. Boulder's smart grid includes updated substations, transformers, and feeders. The Boulder smart grid uses a fiber-optic loop that encircles the city [21]. This network allows for households and utilities to communicate with each other. The Boulder smart grid also has the ability to reroute power during an outage. Participating customers in Boulder have in-home smart meters such as smart thermostats and smart plugs. These smart meters communicate with the smart grid network and allow customers to access their usage information online. Consumers will soon be able to program their appliances online and choose which energy source to use [20]. In addition to in-home monitoring, Xcel Energy has chosen a few homes to act as small power plants [22]. Only a few homes were chosen in this experiment. One example of the smart grid technologies being placed in these test homes is the placement of solar panels on the roof. The panels are connected to a battery that stores energy for later use. This provides backup energy for the consumer's homes and for the grid as well [22].

1.3.2 Austin, TX

Austin, Texas, also started its smart grid project in December 2008. This project is called the Pecan Street Project [23]. This project is run by Austin

Energy and involves several high-tech companies, including IBM, Semiconductor, GridPoint, and GE Energy. This project was able to develop quickly because Texas is the only state in the country that has its own power grid. This bypassed the federal approval process that other smart grid projects in the country have to undergo. This process normally takes several years [23]. Another reason Austin was a prime candidate for such an ambitious project is the fact the city has shown a devotion to clean energy. Austin currently operates the nation's largest "green power program." Austin is also home to the University of Texas, which runs the Austin Technology Incubator and Clean Energy Incubator. These are both major sources for green energy research and commercialization [23]. Many professors from the University of Texas have played active roles in the Pecan Street Project. Another unique feature of the Pecan Street Project is that it is city owned. The board of directors for Austin Energy is the Austin City Council. This allows for Austin Energy to make decisions based on their benefits to the community, giving them a "civic conscience" [23]. Austin Energy chose the Mueller development in Austin to execute the smart grid project. Mueller is a 700-acre development on the former site of the Robert Mueller Municipal Airport. The Mueller community is comprised of new, green, efficient homes and businesses that run off of an on-site power plant [24]. The first phase of the project, entitled Smart Grid 1.0, is a 440-square-mile system that includes over 500,000 devices and involves roughly 100 terabytes of data [25]. This smart grid includes approximately 1 million consumers and 43,000 businesses [25]. The second phase of the smart grid project is called Smart Grid 2.0. This project is already in process. Smart Grid 2.0 has already placed over 80,000 smart thermostats in homes and businesses. Smart Grid 2.0 will be able to self-heal and interact with consumers. This project also plans to be able to store energy and charge electric vehicles [25].

1.3.3 Ontario, Canada

Canada's Hydro One in Ontario operates smart grid technology on a much larger scale than Boulder, CO, and Austin, TX. Hydro One's goal is to create an intelligent communications network to lower costs and raise efficiency. The Hydro One project was started in response to the energy crisis facing Ontario. By 2025, Ontario will have to replace 80% of its current grid system [26]. Hydro One will rebuild the Ontario grid by utilizing smart grid technology. Hydro One will use a two-way self-healing mesh radio network to communicate with devices in the network. These devices include in-home meters. Hydro One had installed over 700,000 meters by December 2008 and plans to install 1.3 million meters by the end of 2010 [26].

1.3.4 Italy

Italy has been a forerunner in implementing smart grid technology. The project was begun in 2001 by Enel, Italy's largest power company. In 2006, Italy made it mandatory for all electricity providers to use smart meters [27]. Since then, 85% of Italian homes have Enel smart meters [27]. Enel designed and developed all of its own smart meters. Enel uses these meters to relay usage information back to its central office and to offer real-time pricing to customers. Enel reportedly has saved $750 million annually since the implementation of the meters [28]. These meters have been successful in reducing costs for consumers as well.

1.4 The Future of Our Grid Systems

The grid system of the future will be vastly different than our current system. The future grid system will still be based on our current electrical infrastructure. It will use many of our existing grid technologies, such as power lines, transformers, and substations. These technologies will be incorporated with new discoveries to build an intelligent grid system. This grid system will be a more efficient, safer, and stable grid network. It will be a fully automated system that will constantly control and monitor power flow at all times. This information will be accessible not only to power plants, but to consumers and businesses as well. This will ultimately make for a more intelligent consumer, which will in turn reduce the overall cost of electricity. This new, efficient system will allow for and encourage economic growth. The future grid system will be comprised of many small, local networks that are connected to a large, national network. Appliances in the homes of consumers will be connected to their local networks. All of these different levels of the smart grid will work together seamlessly to monitor the network in real time. Consumers will also be able to buy back into the network by using plug-in hybrid vehicles. The vehicles will provide energy to the network when they are not in use. New jobs will be created by the new companies that produce the necessary technology for the smart grid that will be needed, and other businesses will not be held back by power constraints. The smart grid of the future will maximize efficiency while reducing the harm done to the environment by using more effective methods of energy transferring and incorporating renewable energy resources. Wind, solar, and hydroelectricity will be connected all across the United States. Our future grid system will use high-transmission cables to connect the East and West Coasts of the United States. These cables will be connected to Canada and Mexico as well. By connecting both coasts and the surrounding regions, the grid system will become more efficient. This interconnected system will reduce line losses tremendously. Many different renewable resources across the United States will be interconnected, providing additional power throughout the grid. Linking the East and West coasts will connect many power storage facilities, which will aid in supplying emergency power across the grid. Wind, solar, and hydroelectricity will be seamlessly connected all across the United States.

1.4.1 Path Forward

The implementation of smart grids is still in the distant future. In order to bring the smart grid vision to a reality, many important steps must be taken. Since the smart grid will consist mostly of two-way communications, cyber security must be addressed. The smart grid will use in-home monitoring, which causes concerns about privacy for many consumers. Political funding and support are also necessary for the implementation of the grid. Additional research is also required for the modernization of our grid system. Once we overcome these obstacles, we can begin building the grid of the future.

1.4.2 Cyber Security

One major concern many people have about the smart grid is the issue of cyber security. The future grid system will be able to control and regulate electricity throughout the grid. If this powerful system falls into the hands of the wrong person, it would be catastrophic for the country. Terrorist attacks of this type are a major concern. A group developed by the National Institute of Standards and Technology (NIST), named the Smart Grid Cyber Security Coordination Task Group (CSCTG), plans to address this issue [33]. The cyber security of the grid is very complex. It encompasses every part of the smart grid, from utilities and power stations to small area networks.

1.4.3 Consumer Privacy

The increase of data being transferred between the consumer and the utility has raised concerns about privacy. Utilities will be able to read customers meters remotely. This has many consumers worried about what information these utilities will actually use. Data from electricity usage can tell many things about what an individual is doing in their home. It can tell whether the person is actually home and when they are awake or asleep. High electricity use can also be an indicator of having expensive electronics and, in the wrong hands, this information could potentially make someone a target for criminals. Smart devices such as smart meters and smart appliances have also been an area of concern for consumers. These devices can be scheduled to work during off-peak hours when electricity is the cheapest. The consumer sets these presets, while the utility controls the power supplied. Information about an individual's habits can be derived from these appliances. The utility could potentially tell when you run your dishwasher, when you shower, and when you watch TV. It can also tell which rooms you use during the day and which rooms you use the most. If this information is not secured, it could lead to identity theft and third-party surveillance in people's homes. Despite the potential negatives of storage using this information, it still remains a vital part of the smart grid. Utilities will use this information to determine load requirements and improve the grid's efficiency. Steps must be taken by utilities to secure this information. New privacy laws are being researched and developed by organizations such as NIST. These groups are investigating ways to make personal data collected by the grid anonymous and secure.

1.4.4 Political Funding and Support

In order to revolutionize our grid system, support is needed from investors and politicians. So far, President Obama has been an advocate of smart grid technology. He has clearly outlined our need for a new grid system in his energy plan. He has allotted millions of dollars of funding for the Department of Energy to conduct smart grid research. The estimated total amount of money invested in smart grid projects through the Department of Energy will be $8.1 billion. However, before development can begin on a wide-scale smart grid, investors and consumers must be educated more on the topic. Utility companies, along with their investors, need to understand why they have to change their whole infrastructure. They must understand that revolutionizing our grid system is vital to our future.

1.4.5 Current Research

Many research projects are being conducted on the topic of smart grids. Cities like Austin, Texas, Boulder, CO, and others have started small-scale smart grids. Other major research projects are also being developed. These include the Modern Grid Initiative, GridWise, Electrical Power Research Institute's (EPRI) Intelligrid program, and many others. Research on the new advanced components for the smart grid is also being conducted.

The Modern Grid Initiative is a program started by the Department of Energy that is designed to "accelerate grid modernization in the United States" [34]. Recently, this program has renamed itself the Smart Grid Implementation Strategy [34]. Some goals of this program include educating and developing relations among utilities, consumers, policy makers, and other important figures in the implementation of the smart grid. The Modern Grid Initiative also will provide "tools, materials, and expertise" to aid in the development of the smart grid [34]. GridWise was founded in 2003. GridWise, much like the Modern Grid Initiative, plans to educate stakeholders on the importance of developing a nationwide smart grid [35]. On April 21, 2010, GridWise and IEEE Power and Energy Society signed an agreement to work together on the development of smart grid technology [36]. EPRI's Intelligrid program focuses on providing and developing the technology necessary to implement the smart grid. So far, the Intelligrid program has provided a number of utilities with their own smart grid technology. These technologies include "advanced metering, distribution automation, and demand response" [37].

1.5 Conclusion

In order for our country to continue to expand successfully, changes to the grid system must be made. The current grid system will not be able to provide the energy we need in the upcoming decade. This new grid system will reduce our dependence on fossil fuels, be more reliant and efficient than our current grid system, and be a more secure way of transferring power. The smart grid is an investment in our future, and it must be developed in order for our country to survive.

Acknowledgment

This work is supported in part by the U.S. National Science Foundation (NSF) under grant CCF-0829827, CNS-0716211, CNS-0737325, and CNS-1059265.

References

1. Grid 2030: A National Vision for Electricity's Second 100 Years. U.S. Department of Energy, Office of Electric Transmission and Distribution. (2003). Web. 14 April 2010.
2. Battaglini, A., J. Lilliestam, C. Bals, and A. Haas. The SuperSmart Grid. European Climate Forum. (2008). Web. 14 April 2010.
3. Holland, S. Obama Announces $3.4 Billion in Grants for Smart Grid | Reuters. *Business & Financial News*, Breaking US & International News | Reuters.com. (27 October 2009). Web. 21 April 2010. http://www.reuters.com/article/idUSTRE59Q1AC20091027.
4. Brown, R. E. Impact of Smart Grid on Distribution System Design. Power and Energy Society General Meeting—Conversion and Delivery of Electrical Energy in the 21st Century, 2008 IEEE. (2008). Print.
5. The Smart Grid: An Introduction. U.S. Department of Energy. (2008). Web. 14 April 2010. http://www.oe.energy.gov/DocumentsandMedia/DOE_SG_Book_Single_Pages.pdf.

6. Mazza, P. Powering Up the Smart Grid: A Northwest Initiative for Job Creation, Energy Security and Clean, Affordable Electricity. Climate Solutions. (April 27 2005). Web. 14 April 2010.

7. Superconductors Play Vital Role in the Smart Grid. NanoMarkets. Web. 20 April 2010. http://www.nanomarkets.net/perspectives/articles.cfm?articleID=216&referrer=AZOMDOTCOM.

8. Superconducting & Solid-State Power Equipment: Plugging America into the Future of Power. U.S. Department of Energy. (6 November 2009). Web. 20 April 2010. http://www.oe.energy.gov/hts_fcl_110609.pdf.

9. Smart Grid Infastructure. American Superconductor Corporation. (29 May 2009). Web. 20 April 2010. http://www.amsc.com/pdf/SMARTGRID_BRO_0509_FINAL.pdf.

10. Fehrenbacher, K. FAQ: Energy Storage for the Smart Grid. Earth2Tech. The GigaOM Network. (13 May 2009). Web. 21 April 2010. http://earth2tech.com/2009/05/13/faq-energy-storage-for-the-smart-grid/.

11. Accelerating the Use of Demand Response and Smart Grid Technologies Is an Essential Part of the Solution to America's Energy, Economic and Environmental Problems. Demand Response and Smart Grid Coalition. (2008). Web. 14 April 2010. http://www.drsgcoalition.org/.../DRSG_Policy_Recommendations_to_Accelerate_DR_and_Smart_Grid-2008-11-24.pdf.

12. NIST Framework and Roadmap for Smart Grid Interoperability Standards Release 1.0. National Institute of Standards and Technology. (2009). Print.

13. Smart Grid System Report. U.S. Department of Energy. (2009). Web. 14 April 2010. http://www.oe.energy.gov/DocumentsandMedia/SGSRMain_090707_lowres.pdf.

14. The Smart Grid: An Introduction. U.S. Department of Energy. (2008). Web. 14 April 2010. http://www.oe.energy.gov/DocumentsandMedia/DOE_SG_Book_Single_Pages.pdf.

15. What the Smart Grid Means to Americans. U.S. Department of Energy. (2008). Web. 14 April 2010. http://www.oe.energy.gov/DocumentsandMedia/ConsumerAdvocates.pdf.

16. Superconducting & Solid-state Power Equipment: Plugging America into the Future of Power. U.S. Department of Energy. (6 November 2009). Web. 20 April 2010. http://www.oe.energy.gov/hts_fcl_110609.pdf.

17. Smart Grid Infastructure. American Superconductor Corporation. (29 May 2009). Web. 20 April 2010. http://www.amsc.com/pdf/SMARTGRID_BRO_0509_FINAL.pdf.

18. Plug in to Materials Trends for Smart Grid Applications. NanoMarkets. Web. 28 April 2010. http://www.nanomarkets.net/perspectives/articles.cfm?articleID=183&referrer=AZOMDOTCOM.

19. Pumped Hydro. Electricity Storage Association. ESA. (April 2009). Web. 28 April. 2010. http://www.electricitystorage.org/site/technologies/pumped_hydro/.

20. Frequently Asked Questions. SmartGridCity. Xcel Energy. Web. 28 April 2010. http://smartgridcity.xcelenergy.com/learn/frequently-asked-questions.asp.

21. Welcome to Smart Grid City, Colorado. High Country News. Web. 28 April 2010. http://www.hcn.org/articles/17704.

22. Simon, S. The More You Know ...—WSJ.com. Business News & Financial News—The Wall Street Journal—WSJ.com. (9 February 2009). Web. 5 May 2010. http://online.wsj.com/article/SB123378462447149239.html.

23. Working Group Recommendations. Pecan Tree Project, Inc. (March 2010). Web. 4 May 2010. http://pecanstreetprojectaustin.org/wp-content/uploads/2010/03/Pecan_Street_Final_Report_March_2010.pdf.

24. Thinking Green. Mueller Austin. Web. 4 May 2010. http://muelleraustin.com/green/.

25. Austin Energy Smart Grid Program. Austin Energy. Web. 5 May 2010. http://www.austinenergy.com/About%20Us/Company%20Profile/smartGrid/index.htm.

26. The Hydro One Smart Network. SmartGridNews.com. Trilliant. (31 March 2009). Web. 4 May 2010. http://www.smartgridnews.com/artman/uploads/1/Hydro_One_Case_Study_012209.pdf.

27. Jones, D. Italy: Smart Grid World Leaders. *Energy Industry News Europe | GDS Publishing*. (18 November 2009). Web. 5 May 2010. http://www.ngpowereu.com/news/italy-smart-grid/.

28. Scott, M. How Italy Beat the World to a Smarter Grid. *BusinessWeek*. Bloomberg. NY (16 November 2009). Web. 4 May 2010. http://www.businessweek.com/globalbiz/content/nov2009/gb20091116_319929.htm.

29. Lindley, D. Smart Grids: The Energy Storage Problem: *Nature News*. Nature Publishing Group. (6 January 2010). Web. 12 May 2010. http://www.nature.com/news/2010/100106/full/463018a.html.

30. Smith W., The Role of Fuel Cells in Energy Storage. *Journal of Power Sources* 86(1–2): 74–83, 2000. DOI: 10.1016/S0378-7753(99)00485-1.

31. Batteries and Ultra-Capacitors for the Smart Grid: Market Opportunities 2009–2010. Rep. NanoMarkets. (August 2009). Web. 12 May 2010. http://www.flextech.org/documents/nanomarkets_reports/SmartGrid_ES.pdf.

32. IdaTech, Cascadiant Deploy Fuel Cells with Hutchison Telecom in Indonesia. Renewable Energy Focus. Elsevier Ltd. (15 April 2010). Web. 12 May 2010. http://www.renewableenergyfocus.com/view/8758/idatech-cascadiant-deploy-fuel-cells-with-hutchison-telecom-in-indonesia/.

33. Lee, A., and T. Brewer. The Cyber Security Coordination Task Group. Smart Grid Cyber Security Strategy and Requirements. (2009). Web. 26 May 2010. http://csrc.nist.gov/publications/drafts/nistir-7628/draft-nistir-7628.pdf.

34. The NETL Smart Grid Implementation Strategy (SGIS). DOE—National Energy Technology Laboratory: Home Page. Web. 15 June 2010. http://www.netl.doe.gov/smartgrid.

35. GridWise® Alliance: About the Alliance. GridWise® Alliance: Welcome. Web. 15 June 2010. http://www.gridwise.org/gridwisealli_about.asp.

36. GridWise® Alliance and IEEE Power & Energy Society Enter into Collaboration Agreement. GridWise Alliance. (21 April 2010). Web. 15 June 2010. http://www.gridwise.org/uploads/news/GWA_IEEE-PES_signMOU.pdf.

37. IntelliGrid. Electric Power Research Institute. Web. 15 June 2010. http://intelligrid.epri.com/default.asp.

38. http://www.worldenergy.org/documents/p001546.pdf.

Chapter 2

Distributed Algorithms for Demand Management and Grid Stability in Smart Grids

Monica Navarro, Lorenza Giupponi, Christian Ibars,
David Gregoratti, and Javier Matamoros

Contents

Smart grids will incorporate an array of new technologies making electric grids greener, safer, and more efficient. Demand management is a fundamental aspect in electric grids, which will gain relevance when automated metering and advanced communications infrastructures between different elements of the grid are in place. In this chapter, we address techniques that will enable demand management for small consumers and residential users. We distinguish between direct load control techniques and pricing policies. In the former, users agree to transfer control of their electricity consumption to the utility, while pricing policies provide incentives for users to manage their demand according to a global optimization criterion. In such a scenario, game theory provides a set of techniques that allow us to model the demand management problem and obtain a satisfactory equilibrium solution. In particular, we model demand management as a congestion game, which possesses at least one Nash equilibrium. Furthermore, congestion games allow the characterization of constraints on the distribution grid.

2.1 Smart Grid Elements

Nowadays power grids are complex systems that have been growing for longer than a century. Nontrivial operations allow large amounts of energy to be transported from production centrals to area substations, where they are distributed to users. To keep the system running properly (what is called the *normal secure state* [1]), production utilities and distribution companies must have reasonable control of the grid state and react fast to any event that may alter the equilibrium. Recently, an increased frequency of grid failures and wide-scale blackouts (US and Canada, 2003 [2,3]; Italy, 2003 [4]; Java, 2005 and Bali, 2009 [5]; and Brazil and Paraguay, [6] just to cite a few) has shown the need for improving, automating and speeding up those control procedures. Additionally, an increased consciousness for environmental problems is motivating the introduction of new solutions to improve grid efficiency, reduce energy losses, and supplement classic carbon/oil-based production plants with modern low-emission energy sources. All these considerations are pushing toward an essential renovation of the global power grid [7–11].

In spite of local specificities, power grids mainly consist of four basic parts: generation, transmission network, distribution network, and customers. High-voltage transmission lines bring electricity from generation plants to distribution substations. Here, electricity is "step down" converted to medium and low voltage to be delivered to customers [12]. As we have already mentioned, future smart grids will surely have to improve generation efficiency and control algorithms for the transmission network. However, it is probably in the distribution network to the final user where the most significant changes will be noticed. In particular, in this chapter we will focus on some approaches aimed at a better match between energy offer and demand (we will see that the target is to achieve uniform peak-free load over time). As a side effect, higher control on the demand and the consequent energy allocation will also allow management of energy coming from different sources, including local generators and storage systems. For this reason, we start by describing briefly the impact of these two technologies in smart grids.

2.1.1 *Distributed Energy Resources*

Today's power grid is designed to operate in a centralized fashion. Namely, the vast majority of the power delivered to the customers is generated in a few, very large capacity plants. This approach has several shortcomings. First, energy is generated far from the loads, thus entailing significant transportation losses. Second, the dependence of the whole grid on few generation plants causes major reliability issues. In addition, grid scalability and evolution require complex, expensive actions at the central level.

To cope with these limitations, distributed energy generation (or, simply, distributed generation (DG)) is a concept that has attracted considerable attention in the last years. The future vision of the electric grid is to build smaller energy generation plants at, or close to, the customer side. The result will be a number of microgrids, semiautonomous systems connected in a dynamic distributed fashion to the main utility grid [13]. Clearly, this new architecture will need an upgrade of the transportation and distribution network. With that aim, the IEEE group has started to develop the IEEE Standard P1547.4 [14], which deals with the design, operation, and interconnection of distributed energy resources (DER) systems. In the following, we summarize the main advantages of microgrids and DER systems:

Reliability: Microgrids will be crucial against grid blackouts since they are expected to cover part or all of the needs of the customers attached to them. For instance, a microgrid can be islanded from the main grid when the latter cannot support the current demand or deteriorates the power quality of the microgrid.

Efficiency and sustainability: Energy will be generated near the customers. This will reduce losses in the transport network. Besides, the relatively low capacity of distributed generators will allow the deployment of, e.g., low gas emissions turbines and an increase in the penetration of renewable energy sources. As a result, CO_2 emissions will be reduced when compared to conventional generation techniques. Combined heat and power (CHP) systems is another technology that will probably evolve with DG and microgrids. The idea is to employ the heat generators dissipate when producing energy for other purposes, such as domestic heating or hot water.

Ancillary services: The Federal Energy Regulatory Commission (FERC Order 888-A, April 1996) defines the ancillary services as "those services that are necessary to support the transmission of capacity and energy from resources to loads while maintaining reliable operation of the Transmission Service Provider's transmission system in accordance with good utility practice." One of the main purposes of a microgrid will be to offer power premium capabilities and ancillary services; for more detailed information the interested reader is referred to Kueck et al. [15].

Scalability and investment deferral: It is foreseen that the increasing energy demand will not be served by the current centralized energy plants. Clearly, the grid could be expanded by building new centralized power plants to meet these needs, but unfortunately, this might not be economically feasible or sustainable. In contrast, with distributed energy generation systems, energy will be generated close to loads, which will provide means to increase energy generation locally and meet the increasing demand in a more efficient manner.

Flexible market: According to the microgrid infrastructure, customers will have the flexibility to consume energy either from the grid or from the microgrid, depending on the power quality or price. Besides, microgrids will have the possibility to obtain higher revenues in case of excess of generation or low demand by selling the excess of energy to other microgrids or directly to the main grid.

2.1.2 Distributed Energy Storage

Energy storage has been identified as one of the cornerstones of the future power grid [16]. Future developments will reduce costs in energy storage systems, which will entail an affordable massive and distributed deployment. This will allow a more sophisticated and efficient control of the grid: higher stability and power quality, higher storage reserves in case of generator failure, etc.

Despite their significant number, energy storage technologies can be categorized into two main groups according to the application they are designed for [17]: *power applications* and *energy applications*. Power applications are those applications requiring a high power with a short discharging time. On the contrary, energy applications require a larger discharging time (usually from several minutes to hours) in order to deliver large amounts of energy. The spectrum of the different energy storage technologies is depicted in Figure 2.1. As it can be observed, within the group of power applications are flywheels and electrochemical double-layer (ECDL) capacitors, whereas pumped hydro and compressed air energy storage (CAES) are more appropriate for energy applications (interested readers are referred to, e.g., Roberts [17] or Johnson et al. [12]).

In the current power grid, energy storage supports the main grid with a number of applications and ancillary services [18]. Obviously, improving these features may raise the efficiency of the grid and, in turn, defer investments in new generation units. Next, we summarize the main applications and ancillary services provided by distributed storage systems.

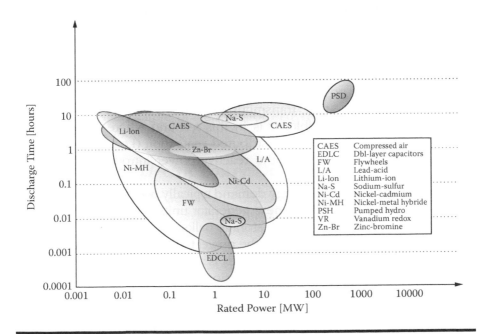

Figure 2.1   Discharging time vs. rated power for different energy storage technologies [17].

is one of the main causes of power supply failure. The electric power grid is typically overdimensioned in order to cope with the demand of few peak hours in a year span. For instance, in Spain about 4,000 MW is required to attend 300 hours of peak consumption per year [27]. The demand also varies significantly within the day, with aggregated national power consumptions varying from 22 MW at 4 a.m. to a 35 MW peak value at 10 p.m. [27]. Despite the fact that demand side management (DSM) programs were originally conceived to shed the aggregated demand peak values, smart grid infrastructures are envisioned to facilitate the design of a modernized load management to deliver power in optimal ways, and respond to a wide range of power demand conditions.

Next, we address the common terminology used in the literature to describe the wide range of programs that intend to influence the electric power demand and its usage patterns. DSM and demand response (DR) are the terms used by electric utilities and the research community to refer to those programs. Sometimes they are used indistinctively. However, there are formal differences. DSM terminology was coined under a broader scope than DR. DSM programs refer to the tools and mechanisms that influence the customer's use of energy. That is, how much and when is electric power being consumed by the end users? DSM programs include different actions taken by the utility to modify or influence the retail customer use of electricity, with the purpose of reducing the individual user's demand, change its use in time, make a more efficient use of it, reduce the aggregated peak load, etc.

DSM may be classified in two broad categories:

Reduce consumption: Actions under this category aim at reducing the consumption of individual end customers. We find actions that improve energy efficiency, raise energy saving awareness, manage energy storage, enable and exploit energy conservation, etc.

Shift consumption: In this case, DSM actions intend to vary the time of consumption, from peak to valley hours, or to delay the load in case of a contingency. More specifically, by:

■ Filling valleys, which may involve making use of storage technologies, pumping stations, allocating new forecast demand for electric supply (charge of electric vehicles).

■ Reducing consumption during system peak hours, which includes pricing incentives, direct load control (including interruption of service), or automatic load management, among others.

Thus, DSM initiatives are intended to modify the consumer demand pattern, with the aim of achieving not only net energy savings but a

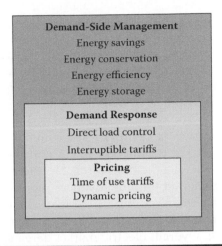

Figure 2.2   Demand side management actions/programs.

more efficient use of the energy itself. A broad classification of DSM is summarized in Figure 2.2.

In parallel to the DSM actions taken by utilities and customers to adopt these measures over the electricity demand, further technical developments of various natures are required in order to exploit the full potential of DSM. To cite a few:

Energy-efficient end use devices: It is important that besides educating the user for responsible energy usage, the appliances are designed under high-energy-efficiency requirements.

Additional electric equipment: Deployment and system integration of additional equipment that allows advanced load shaping mechanisms, for instance, user's generation and storage capabilities.

Communication-enabled devices and infrastructure: Communications between end user and control management are needed, e.g., new devices such as advanced metering infrastructure, or devices that realize simple user control over their own energy usage, or design of the proper interfaces with appliances.

In Sections 2.3 and 2.4 the tools for demand side management and demand response are addressed in more detail. In particular, in Section 2.3 we focus on policies where loads are directly controlled by the utility, whereas Section 2.4 focuses on pricing, and game theory tools for decentralized and dynamic pricing deployment. In the sequel, we make indistinctive use of DSM and DR terminology.

2.2 Electricity Markets

Wholesale electricity markets have experienced substantial changes in the recent past. Traditionally, electricity was delivered by regional, state-owned electric companies. These were characterized by a high degree of vertical integration, consisting in the generation, distribution, and retail commercialization of electricity. The reasons for a regulated electricity market were the particular characteristics of electricity as a commodity, which is difficult to be stored and needs to be produced and consumed in real time, coupled with the need to provide electricity in a reliable and robust fashion.

Technological and market trading advances prompted the deregulation of electric markets in several countries. Nordic countries established the Nord Pool as a cross-national, liberalized electricity market, and other countries followed suit [28]. Overall, electric market deregulation has produced a healthy competitive environment, but malfunctions such as the 2000 California electricity crisis highlight the importance of a properly designed market that is adapted to the technical limitations of generation, distribution, and commercialization of electricity. In particular, generation and transport capacity need to be properly overdimensioned in order to avoid excessive price volatility in case of demand spikes, and the market needs to ensure that all parties involved have the possibility to make a profit.

It is worth noting that a deregulated market provides price incentives for demand management by consumers. A consumer that is able to focus its demand on valley periods where supply is abundant will be able to secure lower prices. Therefore, the wholesale market provides the basic incentives for electricity retailers, or utilities, to apply demand management techniques at the retail level, which are the object of this chapter. In the following sections we will argue that pricing is also a valuable tool for demand management at the retail level. For now, in this section we describe the basic structure of a deregulated wholesale electricity market.

2.2.1 Market Agents

The basic agents of the market are producers, consumers, and the market operator. Electricity producers constitute the supply side of electric markets. The array of technologies used to generate electricity is vast, and varies widely in the amounts of power produced as well as the capability to produce electricity on demand (see Table 2.1). An important characteristic is the availability of a supplier to produce a certain amount of power on demand, which leads to the characterization of dispatchable vs. non-dispatchable sources. A dispatchable source is able to match a predefined production schedule, whereas the production of a nondispatchable source will depend on uncontrollable factors, such as wind, irradiated sun power, etc. Dispatchable sources are therefore more suitable for long-term

Table 2.1   Availability of Energy Sources

Type	Volume	Availability	Cost (operational/investment)	Carbon Footprint
Hydro	Med./high	Seasonal	Low/Med.	None
Coal	Med./high	Dispatchable	Med./Med.	High
Nuclear	High	Dispatchable	Med./High	Med.
Gas	Med.	Dispatchable	High/Med.	High
Wind	Low/Med.	Nondispatchable	Low/High	None
Solar	Low	Nondispatchable	Low/High	None

trading, and they are also reliable reserve power sources. On the other hand, nondispatchable sources may be unable to meet a prespecified demand and will need a dispatchable source as backup. As we will see, these two types of power generation must be traded in different markets.

In the demand side of the market, we distinguish between retailers and large consumers. Retailers buy electricity in the wholesale market and distribute it to small consumers. Typically, retailers have no generation capability. Large consumers are end users of electricity that participate in the wholesale market. As in the supply side, consumers of electricity may have several degrees of flexibility in their electricity consumption. A consumer that is able to schedule its demand should be able to adapt to supply and obtain lower energy prices. Likewise, retailers, as consumers in the wholesale market, also benefit from the ability to schedule their demand according to market prices.

The market operator (MO) is responsible for the operation of the energy transactions. Based on the received bids, it will determine the prices, quantities, and schedules of electricity delivery. We refer to this process as *market clearing*. Figure 2.3 shows the interrelations between different wholesale market agents. The retail market is represented by the interaction between retailers and end users, regulated through service contracts.

MO may also perform services critical to electricity exchange, such as transmission system operator (TSO), and also provision of ancillary services that guarantee the integrity of the system. Other functions of the MO include managing congestion and ensuring that sufficient generation capacity is readily available.

2.2.2 Market Operation

In the electric market, producers submit production bids while consumers and retailers submit consumption bids. For example, a producer may submit

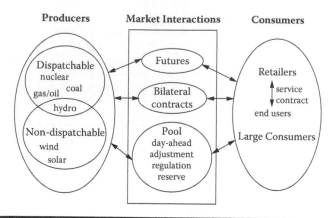

Figure 2.3 Market agents and their interrelations in a deregulated electricity market.

a bid to produce A MWh at a price p_a ($/MWh), an additional B MWh at a price p_b, and so on. Likewise, a consumer will bid to buy C MWh at a price p_c, an additional D MWh at price p_d, etc. The MO receives all bids and allocates production and consumption accordingly.

The physical nature of electricity does not allow instant pricing and delivery of electricity as would be possible with other commodities. Contracts must be negotiated some time in advance so that production can be planned and scheduled. Since the precise demand is not known, such contracts are based on a demand forecast. However, demand must be met precisely to avoid service outages. In addition, nondispatchable sources may not meet their production forecast, so a compensating source must kick in. In order to meet real-time demand and at the same time facilitate planning, electricity trading is organized in a sequence of exchanges with decreasing time to delivery. This mechanism allows market agents to plan their production in advance, and at the same time ensures the stability of the electric grid by matching supply and demand in real time. In the following we describe such exchanges, ordered as a function of time to delivery.

Bilateral contracts and futures: Two parties may agree on an electricity delivery schedule for weeks, months, or years ahead under a bilateral contract. The contract may include different terms of several years, down to specific hour or half-hour intervals. This market is used to trade production and consumption that can be forecast very reliably a long time before delivery.

Next-day trading: Some demand may not be forecast early in advance and cannot be traded as futures or long-term contracts. Next-day trading offers the possibility to trade production and consumption one day ahead of delivery. Parties submit their bids for electricity for the next day, specified at time intervals (typically 1 hour or half-hour) before a certain time. The MO collects these bids and clears the market, allocating production and consumption. Next-day trading is the most important short-term market by volume.

Adjustment markets: Next-day trading is based on forecasts that may be refined as time-to-delivery approaches. Adjustment markets provide a mechanism to submit supply/demand bids only a few hours prior to delivery. A nondispatchable producer may use this market to compensate for errors in its forecast of energy production. In the adjustment market, suppliers and consumers submit offers (increase generation/reduce demand) and bids (reduce generation/increase demand) with respect to the allocations in next-day trading.

Regulation markets: The need to ensure precise demand and supply match requires real-time delivery of energy. Regulation markets are used to provide energy in an *as-needed* basis.

Reserve markets: Besides matching demand, additional supply must be available to react to failures in electricity generation or transport. The reserve market is used to trade reserve capacity (which is on standby and ready to deliver on a very short notice) to compensate for a certain amount of generating capacity going offline.

Electricity producers will participate in different markets depending on the nature of the generating technology. Sources that are able to meet a certain demand on short notice may participate in adjustment, regulation, and reserve markets, while sources with a slower adaptation or seasonal availability must sell their output through long-term contracts. The timeline of different trading scenarios is shown in Figure 2.4. An important parameter is the time at which information related to trades needs to be relayed to parties, called *gate closure*.

The MO is responsible for clearing the market and determining the prices and quantities of electricity sold at a given time slot. The output needs to be optimal (in the sense of maximizing social welfare) and at the same time take into account constraints such as fairness and robustness. In addition, in situations where the distribution grid imposes a limitation, the solution may include a geographical component; i.e., prices are given for generation/consumption at specific locations. The solutions

Figure 2.4 **Timeline of market activity with respect to delivery time T.**

implemented are in general different for different electricity markets. The MO needs to make decisions with incomplete information regarding actual supply (if renewable sources are involved) and demand. In addition, decisions are taken at several points in time where additional information may become available. Such problems are well modeled as multistage stochastic optimization problems [29]. Stochastic optimization provides a solution framework when random variables, rather than deterministic values, appear in the objective function or constraints of the problem. In addition, multistage problems may be used to model several decision points with a different level of uncertainty (i.e., the outcome of random variables is known or not).

An important aspect in the solution is security. Since markets are cleared ahead of delivery time, unforeseen events, such as power plant failures, or transmission line outages, may take place, which limit the availability of electricity below demand. Security refers to the capacity of the solution to withstand a certain number of such disturbances occurring simultaneously. This is accomplished by properly scheduling reserve capacity.

A deregulated electricity market was made possible with technological developments that modernized the grid during the 80s and 90s of the last century. A new wave of technological developments is making inroads into the grid and will result in a grid that is smart, flexible, and efficient beyond today's limits. Some notable developments include:

Increase in schedulable demand: Smart appliances will facilitate scheduling of energy consumption. In addition, home energy managing devices may be used to schedule appliances and adjust load without user intervention.

More robust communication: Robust and real-time communication will increase reliability and efficiency of the grid.

Increased buffering capability: Distributed storage solutions addressed in Section 2.1 will allow the participation of new agents, such as large plug-in hybrid electric vehicle (PHEV) fleet owners, to participate in electric markets.

On the other hand, renewable energy sources are in general non-dispatchable, which further complicates real-time energy trading. New trading structures will be necessary to take advantage of new technologies, respond to new challenges, and produce a robust competitive environment for energy trading.

Regarding the retail market, end users are faced with little flexibility to deal with their demand. It is foreseen that demand management techniques such as those described in the following sections will help retailers optimize the demand of their consumer base and obtain better prices in wholesale markets.

2.3 Demand Response

Among the mechanisms for control of the electric power consumption, in this section we focus on the set of DR actions that aim to control the customer use of electricity by means of cost and economic incentives.

There exist different classifications of DR programs that are related to different aspects of the grid. It depends on the volume of electric power demand contracted by the customer, which implicitly determines to what segment it is connected to (high, medium, or low voltage) and the ability to participate in the different electric markets (wholesale or retail). The specific DR that one may implement also depends on the metering capabilities deployed in the power grid. With the deployment of advanced metering infrastructure (AMI) or smart meters (SMs) we will see that advanced demand response strategies are being implemented. AMI are electronic devices with sensing and communication capabilities that periodically report the energy consumption and retrieve pricing information. It is this last capability that makes this technology valuable for the future smart grid.

We will distinguish in this section between mechanisms that envisage a direct control of the retailer on the load generated by the customers and other schemes that avoid the direct intervention of the retailer by encouraging load balancing through dynamic pricing policies. In general, pricing policies are designed to charge less per kWh during off-peak periods (typically at night). Although there exists a vast number of contributions to understand the mechanisms of pricing applied to electric DR, their application in real systems varies considerably from country to country, partially due to their specific regulatory constraints.

The DR programs in current power grid systems may be classified in three different categories, depending on the degree of intervention on the loads by the system operator (SO):

Direct load control (DLC): Based on this method, the SO has a direct control over specific loads, e.g., household appliances. This approach requires communication and control mechanisms to actuate on the particular loads by turning them on/off, delaying the operation, etc.

Pricing policies: If the SO is unable to directly intervene on household appliances, it can only provide incentives to make users manage their traffic. As a result, this is an incentive-based approach, which comprises strategies that establish different prices for energy consumption, depending on the time of the day the electric power is actually consumed. These policies are applied at different segments of the grid. For residential users, pricing policies have also been the most extended tool to encourage shifting consumption from peak to valley hours.

Interruptible tariffs: Intermediate solutions are also possible. For example, this program offers cheaper electricity to customers in exchange for reducing their consumption when requested by the utility or the system operator. Notification minimum times are established by contract, as well as the maximum number of service interruptions and maximum duration of the interruption.

Both incentive-based and direct-control-based DR programs are falling short for a successful demand management in energy efficient networks. We have seen in Section 2.2 that markets, and consequently regulation, are evolving in the right direction to provide the necessary tools for more efficient pricing policies. Today, pricing incentives range from simple time-of-use tariffs that only distinguish between day and night tariffs to finer time resolution tariffs for customers with advanced metering capabilities (e.g., half-hour/hourly readings). In the first case, the tariff may be updated as frequently as on a seasonal basis [30], thus requiring very basic communication infrastructure or none at all.[1]

Direct load control (DLC) has been limited so far to large consumers, although some countries managed to offer DLC to residential and small business users. We will argue in the next sections how the smart metering programs that are being deployed today around the world could further extend DR programs to large-scale deployments that include all ranges of customers (from large industrial users to household consumers[2]). On the other hand, DR algorithms would need to make this process automatic and of simple operation for the end user, accommodating both customer priorities and system operator requirements. Next, we review the basic mechanisms of DLC and pricing.

2.3.1 Direct Load Control

Direct load control is a contractual agreement between the electricity provider (utility) and the customer, through which the customer allows the system operator to directly schedule, reduce, or disconnect part of the load, in return for financial incentives. DLC has been in place for decades, mostly applied to industrial and large commercial customers. Recently, these programs have been extended to reach residential and small enterprise customers.

---

[1] In Spain, DR applied to residential customers (below 10 kWh of contracted power) consists of a pricing policy that distinguishes only between day and night tariffs, with day/night intervals spanning an 8/14-hour period, respectively. Starting/ending times are modified according to winter/summer season.

[2] The DR mechanisms, though, would be different since they participate in different markets.

DLC actions require some control capability at the customer side to process control signaling. In the majority of cases, this is enabled by a one-way communication system, from the utility to the customer. The customer initial investment in the load control device, as well as the intrusive aspects that may derive from direct household operation, have prevented residential customers from widely participating in DLC programs. However, the potential benefits of DLC are known to be larger if they reach the residential customers, and not only large industrial and commercial users. This is why improving DLC mechanisms is still considered a valid approach for smart grid systems. In the following, we review the basic operation functionality of DLC programs. Special attention is paid to DLC techniques that consider small customers.

At the commercial and residential levels, the type of loads typically managed by DLC programs have mostly included thermal systems such as heaters, air conditioning units, and lighting systems. The reason being twofold: cooling and heating are one of the largest sources of household energy consumption and have significant tolerance to on/off operation due to the thermal inertia. Currently, white-good manufacturers are also considering extending load control capabilities to appliances such as refrigerators, washing machines, ovens, etc.

Implementing DLC programs for small consumers has been conventionally done by means of an aggregator manager that gathers a critical mass for load control. Small customers are associated to the aggregator, which interfaces with the system operator. The classical approach to manage the aggregated loads is to group appliances of the same type and to control all the appliances in the group in the same way.

The main goal pursued by DLC is to reduce both the cost of energy supply and the peak load (to prevent peak demand does not exceed safety levels that jeopardize the grid operation). In fact, DLC may be defined as a scheduling problem. For that purpose an important aspect for an efficient operation of DLC is the modeling of the loads, at both individual and aggregated levels. In particular, DLC actions need to have an accurate knowledge of the aggregate load behavior, either by means of direct aggregated load models or by accurate models of single loads and good extrapolation for large population of loads. Such degree of accuracy is needed for two reasons, namely, to obtain an accurate load saving estimation (for correct dispatching) and for the optimization algorithm, deriving scheduling of loads.

There is a vast literature on load modeling [31–37] and algorithms to optimize the cost function [38–41]. In Figure 2.5, a generic scheme highlights the main components of the DLC block diagram. Loads are identified not only by the individual customer but by the group type they are associated to. For instance, the DLC actions would schedule all airconditioners associated to a set of customers. The aggregator is the device that actually

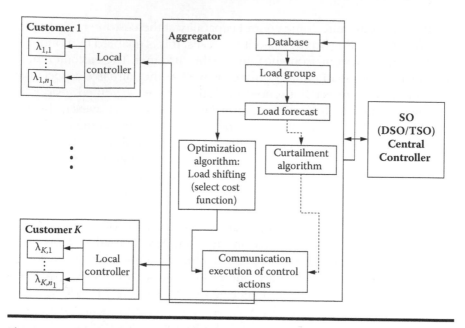

Figure 2.5   Direct load control block diagram.

executes the DLC control actions. It is formed by a database that contains customer information, single load models, pricing information, etc. The database is built and updated by the SO through a bidirectional communication channel. Input signals are then processed, updating customer groups if necessary, and producing a new load forecast value of the loads that need to be rescheduled. We have distinguished between the load shifting and curtailment blocks to stress that both algorithms are run independently. There is an optimization algorithm specifically implemented for rescheduling loads in nominal load-shifting operation, and one for emergency situations when the curtailment algorithm[3] needs to be applied. A set of actions are output by the optimization algorithm, which are executed by means of a communication channel, unidirectional in most cases.[4] At the customer side, a local controller is needed to interpret the DLC actions and actuate directly on the loads by, for instance, switching off the air conditioning or dimming the lights. In the last few years, an effort has been made to provide appliances with better control and communication capabilities that would allow remote operation.

---

[3] *Curtailment* is the technical term used for DSM/DR programs that implement DLC actions to cut down loads during a given period of time.

[4] Not always using analog or digital communication transceivers; it may be executed by means of a telephone call, or by sending a fax to the customers.

Regarding DLC algorithms, different optimization tools have been used in literature. The most widely implemented are based on *linear and dynamic programming tools* [42–45]. But recently, load scheduling algorithms based on *fuzzy logics* [46] and *genetic algorithms* [47] have been proposed for DLC, in particular in the framework of low-voltage consumers. With respect to the optimization criteria, load-shifting algorithms are typically optimized to either meet a predetermined target load curve or minimize the cost of energy. Note that, given the higher cost of electricity in short-term adjustment markets, these two objectives may coincide. Pricing policies may also be included in the model. Some of the recent work toward DSM in smart grids is on optimization algorithms that have been further developed to include two-way communication capabilities [48,49].

Taking all these parameters into consideration, the algorithm outputs the schedule of the appliances' connection in discrete-time intervals and the on/off cycle curtailment.

While it is out of the scope of this chapter to give details on the optimization algorithm, we show the basic components of an optimization algorithm for load shifting (Figures 2.5 and 2.6). Besides defining the cost function, one should carefully define the constraints of the optimization problem. This is illustrated in Algorithm 1, where the load is denoted by

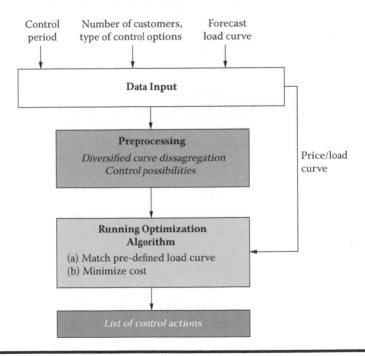

Figure 2.6 Optimization algorithm main blocks.

γ and the time (discrete-time steps) by z. Optimizing the load scheduling requires:

■ Defining the discrete-time intervals of the control period, time-step value, start and end of the control period.

■ Inputting a load forecast curve, jointly with the number of devices controlled in each group, the definition of each group type, etc. A preprocessing is applied before running the optimization algorithm to account for the different criteria, like customer comfort, control possibilities, payback load.

■ Defining the optimization as a minimization problem subject to a series of constraints.

■ Calculating the total load for each group at each time step, which involves the estimation of the load variation. For such purpose, the algorithm typically differentiates between the load variation due to the actions starting before the time step being evaluated and the load variation resulting from the control actions starting at the current time step.

To this end, we shall remark that DLC programs are still being considered a valid mechanism to exploit the full potential of demand side management. Improvements are along the lines of making large-scale deployments a reality by incorporating two-way communications between the utility and residential customers to enable real-time DSM. Along these lines, projects like the Spanish GAD (active demand-side management) [37] have made an effort to develop a communications architecture that enables the implementation of DR based on DLC when smart metering is available at every household.

Algorithm 2.1 : Optimization Algorithm, γ_z

Require: z_0, z_E initial/final step time of the control period, Δt step time duration, $\hat{\gamma}_z$ load forecast, N_k number of devices controlled in group k, k customer type/group.

1. Define cost function as $\min \sum_{z=z_0}^{z_E} \gamma_z$.
2. Define total load at each time step for each group $\gamma_z \doteq \sum_{i=1}^{z} \gamma_i$.
3. Compute total load at each time step z:

$$\gamma_z = \hat{\gamma}_z + \Delta\gamma_z$$

with $\Delta\gamma_z$ being the load variation as a result of the control actions. It is typically divided into two terms: variation due to control actions starting at z, $\Delta\gamma_{sz}$, and starting before z, $\Delta\gamma_{bz}$, respectively:

$$\Delta\gamma_z = \Delta\gamma_{sz} + \Delta\gamma_{bz}$$

2.3.2 Pricing Policies

As already mentioned in previous sections, DR refers to the set of actions and algorithms implemented by utility companies to adjust energy consumption of consumers, in order to more efficiently manage the available energy. One approach is DLC, according to which the utility company can remotely control the energy consumption of appliances in households. The main drawback of this approach is the lack of scalability and the user's privacy. This was discussed in Section 2.3.1.

An alternative is to gradually move intelligence toward the edges of the network and decentralize the decision-making process, which tackles both issues of scalability and privacy. Utilities can intervene and regulate this distributed process by applying smart pricing policies, based on which users are encouraged to individually manage their loads, by either reducing their energy consumption or shifting their consumption from peak hours to less congested hours, thus favoring load balancing. Let us define the following components of a user demand:

Nonessential: Energy consumption can be avoided if the price is high. As an example, heating may be set at a lower temperature if electricity price is unusually high.

Shiftable: Consumption is necessary, but may be scheduled at a different time during the day. An example is a washing machine cycle.

Nonshiftable: This type of demand must be serviced at a specific time. In this case there is no possibility of demand management. Lighting or cooking stoves would generally fall into this category.

For nonessential load, demand is a function of price. Typically, an increase in demand increases the cost of electricity for the utility and results in higher overall prices. This phenomenon is equivalent to congestion. When a user increases its demand, this causes a negative effect on other users as the electricity cost for the utility, and therefore the price it charges its customers, increases. We talk about *externality* when a participant in a market can, without suffering penalty, make choices of variables that adversely affect the utilities of other participants. Increasing demand therefore creates a negative externality to other users. In order to reduce this effect, we may use congestion pricing to *internalize the externality* and make the users pay for it. *Congestion pricing* [50] is a concept aimed at making selfish users pay for the cost their usage imposes on other users, thus forcing them to behave in a way that is socially optimal, which may result in cutting the nonindispensable energy consumption.

In the case of shiftable load, total demand cannot be reduced by pricing incentives; however, pricing can be used to induce desired scheduling patterns that help the utility match the load of its customer base to the electricity obtained in wholesale markets. Matching demand to a particular load has the advantage of increasing predictability of the demand, and also allows the utility to purchase electricity during valley periods where the cost is lower. Therefore, proper demand scheduling represents a clear cost advantage for the utility. The demand scheduling problem is equivalent to the problem of routing traffic in the Internet or in transportation networks, so that we can take advantage of results already obtained in these fields. For example, each time slot in a daily cycle may be seen as a parallel route from source to destination. Scheduling is then equivalent to routing within the set of parallel links. This problem has been extensively studied within the context of game theory. In 1952 Wardrop [51] introduced a traffic game to model the selfish behavior in road networks. Since this concept is also well suited for the analysis of uncoordinated communication networks like the Internet, the model has attracted the interest of theoretical computer scientists over the last 10 years. In Wardrop's model a rate of traffic between each pair of vertices is modeled as a network flow. The resources are the different available paths from the origin to the destination, and the cost (delay) of using a given path increases with the load of that path. Each user chooses the minimum cost route, which leads to an equilibrium point, the Wardrop equilibrium, where all used paths between given origin-destination pairs have equal and minimum latency. Wardrop equilibria are also Nash equilibria, which have the desirable property that, once attained, no player has any incentive to deviate, as it cannot decrease his cost by unilaterally deviating to another path. Wardrop's model assumes an infinite number of users. When we need to consider a finite number of users competing for shared resources, as is the case in a real-world problem, Rosenthal [52] introduced congestion games. As in the Wardrop model, the cost of a given strategy or path increases with the number of users choosing it; therefore this concept can be used as a pricing mechanism to evenly distribute load. In the following section we discuss the applicability of congestion games to demand management problems.

2.4 Network Congestion Games for Demand Side Management

As discussed in the previous section, smart pricing policies and distributed approaches, where the decisions are made locally and directly by the end consumer, have recently taken more relevance. In this scenario, game theory [53] provides a framework to evaluate and design active side management policies [54,55], since it naturally models interactions in distributed decision-making processes. In this section we propose a game theoretic scheme where end customers are the players, the set of strategies is the distribution of the demand across the day, and the cost function the players aim at minimizing is the price they will be charged by the provider [56]. The objective of the game theoretic framework is to obtain a distributed end user energy consumption schedule based on a dynamic pricing strategy. In a similar fashion to Moshenian-Rad et al. [55], electric power demand is distributed over a daily cycle. However, the problem is formulated in terms of a congestion game, which naturally provides a framework to model network competition where the resulting cost is a function of the level of congestion, i.e., the aggregate demand. Congestion games are closely related to potential games [57], and share some remarkable features. In particular, a congestion game is a potential game as it admits a potential function. Finding a solution to a congestion game is equivalent to finding a (local) optimal solution to this potential function. In addition to this, any improvement path is finite and leads to a pure strategy Nash equilibrium. In other words, even though the system is decentralized and all players are selfish, by seeking to optimize their individual objective they end up optimizing a global objective, the potential function, and do so in a finite number of steps regardless of the updating sequence.

With respect to the general framework, we consider a generic smart grid model where users have demand management capabilities. We argue user needs for electric power may be variable in both quantity and time. Such variability may be exploited by the network to schedule demand to optimize network load. For that matter, we envision users equipped with a control device that manages their total electric power demand and which can actuate directly on their own intelligent appliances, storage devices, etc. Internal management is left to the users without direct intervention of the utility. We assume that users know its own total electric demand on a daily basis and decide the exact demand distribution over the time slots spanning a 24-hour period. We assume that this period is divided into 24 one-hour time slots, and refer to a user's demand vector as the value of demand allocated to each time slot. All users participate in a bidding system in the following manner: in an iterative fashion, each user makes its own bid by broadcasting its demand vector to the network. Based on that the network adjusts the price per time slot according to the aggregate demand. Users then may decide to change their consumption priorities by modifying their bids according to the new pricing policy. The process is repeated every time a user updates its bid until convergence is reached. Once convergence is reached (bidding time period), prices per hour and users' demand distribution are fixed, and executed accordingly during the day.

Given that congestion games solve the problem of routing units of load over a general network, they can also be used to solve demand management problems with constraints on the distribution grid. In order to understand this, consider a scenario where a large group of users is connected to the grid through a low-capacity powerline. In this case, demand peaks not only affect overall prices, but may also represent a risk of power outage if the powerline capacity is exceeded. A congestion game may be used to model the distribution grid and will naturally charge higher prices to users behind the congested powerline, contributing to reduce demand peaks and increase overall reliability of the grid. This aspect is also considered in the following sections.

2.4.1 Network Model for Demand and Grid Load Management

Consider an electric grid with a single generation point and multiple customers. In such network, power generation and power demand for each user are time varying, following a daily cycle, as well as other longer or shorter cyclic variations. In our model we assume that electric power is requested and allocated in one-day periods, discretized in N time periods (e.g., 24) called time slots. We model the demand of a user by vector $\mathbf{d}_i = [d_1, \ldots, d_N]$, $i = 1, \ldots, M$, with M being the total number of customers. We map such a system to a network modeled as a directed graph (V, E), connecting the generating point vertex V_0 to the customer vertex, V_1; as shown in Figure 2.7. It is assumed that all customers are directly connected to V_1. V_0 and V_1 are connected with parallel edges $e_{0,1}^1, \ldots, e_{0,1}^N$, representing the different time periods. The power demanded by user i on time slot j flows through edge $e_{0,1}^j$. The cost of such flow depends on the total load in the edge, $x_{e_{0,1}^j}$, and is denoted by $c_{e_{0,1}^j}(x_{e_{0,1}^j})$.

This network model represents power generation only. However, while such a model is useful to manage and optimize power generation in time, it fails to account for other problems, such as peak load in certain segments of the power distribution grid. Overload of certain segments of the electric grid

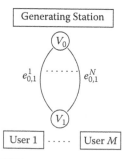

Figure 2.7   Network model for demand management.

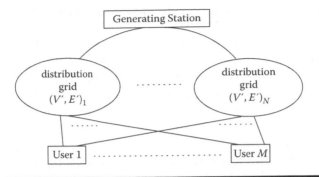

Figure 2.8 **Network model for grid load management. Parallel networks $(V', E')_1$ $\cdots(V', E')_N$ represent the utilization of the power grid during different time slots. This results in N parallel edges for each original network edge.**

is one of the main causes of power failure and blackouts. Since many distribution circuits have been operating close to their operating limits, when one of them fails, neighboring ones must support the total amperage. However, if they are also operating close to their maximum capacity, they will also fail, causing a chain failure [58]. As a result, we extend the demand management model to take into account load constraints in the distribution network, in order to optimize the usage of installed distribution capacity and minimize installed capacity margins, which adds a considerable cost in the distribution grid.

The network model for such a system is shown in Figure 2.8. Here we also assume that electric power is requested and allocated in 1-day periods, discretized in N time slots. In addition, we assume that a model consisting of a directed graph (V', E'), is used to represent the actual distribution network. In the model, vertices may represent the generation point, end users, or intermediate distribution stations. The capacity of power lines is represented by the maximum capacity of the edges. As in the previous model, the cost on the flow along each edge depends on the total load in the edge and is denoted by $c_e(x_e)$.

2.4.2 *Congestion Games*

A congestion game Γ is defined as a tuple $\{M, E, \{S_i\}_{i \in M}, \{c_e\}_{e \in E}\}$, where M is the set of players, E is the set of resources, $S_i \subseteq 2^E$ is the strategy space of player i, and $c_e : \mathbb{N} \to \mathbb{R}$ is a cost function associated with resource $e \in E$. $S = (s_1, \ldots, s_m)$ is a *state of the game*, in which player i chooses strategy $s_i \in S_i$. We assume that players act selfishly and aim at choosing strategies minimizing their individual cost, where the cost c_i is a function of the strategy $s_i \in S_i$ selected by player i, and of the current strategy profile of the other players, which is usually indicated with s_{-i}. The cost function is defined by $c_i(s_i, s_{-i}) = \sum_{e \in s_i} c_e$.

Given any state S, an improvement step of player i is a change of its strategy from s_i to s_i', such that the cost of player i decreases. A classical result from Rosenthal [52] shows that as each user tries to increase his utility, he increases a global function of the allocation. That function is called potential of the game, and it is concave in this case. Thus, as users act selfishly, they increase the potential that then converges to its maximum, which corresponds to the unique Nash equilibrium for the game. By definition [53], a set of strategies for the users is a Nash equilibrium when no user benefits from changing his strategy unilaterally. A potential function $\Phi : S_1 \times \cdots \times S_m \rightarrow \mathbb{R}$ is defined as

$$\Phi(S) = \sum_{e \in E} \sum_{j=1}^{x_e} c_e(j) \tag{2.1}$$

where x_e denotes the number of those players using resource e, i.e., in our model, as previously mentioned, the total load of edge e. Nash equilibria are the only fixed points of the dynamics defined by improvement steps. Hence, the finite improvement property immediately implies the existence of pure Nash equilibria in congestion games.

Games that admit an ordinal potential function, i.e., a potential function with the property that an improvement of an individual player also increases the potential, are called *potential games* [57]:

$$c_i(s_i, s_{-i}) > c_i(s_i', s_{-i}) \Leftrightarrow \Phi(s_i, s_{-i}) > \Phi(s_i', s_{-i}) \tag{2.2}$$

for each player i, strategy profile (s_i, s_{-i}), and strategy s_i'.

Rosenthal's potential function can be shown to be ordinal, so that congestion games can be demonstrated to be potential games. In fact, Monderer and Shapley [57] have also shown that every potential game can be represented in the form of a congestion game. Since every potential game has at least one Nash equilibrium [57], then convergence is guaranteed. It can be seen that the strategy profile for which Φ is minimal is a Nash equilibrium. In particular, if Φ_{min} denotes the set of minima of Φ, and $z \in \Phi_{min}$, then z is a Nash equilibrium of Γ.

A particular class of congestion games is network congestion games, which are described by a directed graph where each player i is characterized by a source V_0 and destination node V_1, and by a set of strategies, which are the possible paths from V_0 to V_1. Every player seeks for a minimum cost path connecting its source with its destination. The cost of each edge depends on the load of that edge, and commonly is assumed to be a non-decreasing function. The game is defined to be symmetric if all the players have the same origin and destination. In this special case, one can compute a Nash equilibrium with the help of a min cost flow algorithm [59].

2.4.3 Demand and Grid Load Management Game

We have modeled in Section 2.4.1 both demand and grid load using a directed graph where cost of a unit load over an edge depends on the total load over that edge. Such problems can be solved in a distributed fashion by defining a noncooperative congestion game among users, which is used to set the price levels for the demand vector of each user. In the following we will focus on a particular kind of congestion game, referred to as a network congestion game, which is also characterized by having a polynomial algorithm to reach convergence to a Nash equilibrium of the game. We take advantage of network congestion games to solve the demand management and the grid load management games. In such a setup:

■ The players are the M users demanding resources during the day.
■ The resources E are the N available edges $e_{0,1}^j$, representing the N time slots, through which the power demanded by user i flows.
■ The strategy of player i is the demand vector d_i.
■ The cost of each resource in E is $c_e = c_{e_{0,1}^j}(x_{e_{0,1}^j})$.

Players define a strategy and update it based on the other players' strategies, until an equilibrium is reached. The Nash equilibrium may be reached if all users update their strategy in order to minimize their cost given a fixed strategy of the other players. Given the network model, such a problem consists in finding the minimum cost flow from the generation point (i.e., V_0) to the user (i.e., V_1). However, link cost is dependent on the total link load, and each user wishes to allocate more than one unit of load. A simple transformation may be used to convert the network graph with load-dependent link costs to a graph with fixed link costs, consisting in splitting each link of capacity K into K unit capacity links with increasing cost $c_e(1), \ldots, c_e(K)$. Once this transformation has been made, the min cost flow problem may be solved using a linear program [60].

The demand management game may be extended to represent and manage load in each link of the electric distribution grid. Under such a model (see Figure 2.8), the congestion game loses the symmetric property, since different users will in general be connected to different points of the distribution grid. Nevertheless, a nonsymmetric congestion game is also guaranteed to have at least one Nash equilibrium, which may be achieved when each user seeks to minimize its cost function, given fixed strategies of the other players. As in the previous section, the cost function minimization problem is the classical min cost flow problem from the generation point to the user.

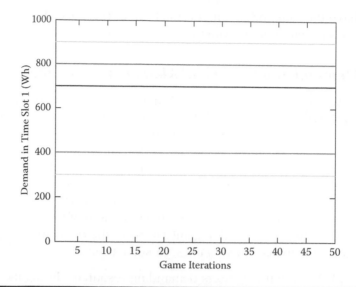

Figure 2.9 **Convergence of user demand over several game iterations, for time slot 1 and for $M = 30$ users.**

2.4.4 Numerical Example

The proposed congestion games for demand management were evaluated through simulation. For the demand management game, $N = 24$ time periods were defined. A symmetric population of $M = 30$ users with three different energy demand profiles was used.

The proposed congestion game was shown to reach an equilibrium solution determining both the user demand vectors and prices paid. Figure 2.9 shows the (almost instant) convergence of the distributed demand management scheme over several game iterations, where the equilibrium point is readily reached. At the equilibrium point, the demand distribution over time slots of all users is shown in Figure 2.10. Figure 2.11 shows the aggregate demand of all users using the demand management scheme and compares it to that of an unmanaged scheme. As it can be seen, the management scheme results in a flat aggregate demand rate, a desirable property from the utility perspective.

2.5 Introduction to More Complex Models

In the two previous sections we have discussed the DSM problem in its simplest form, that is, when a single utility provides energy to all the users. These users are assumed to consume the energy they purchase in real time and no storage is considered. Besides, according to this model, there is no

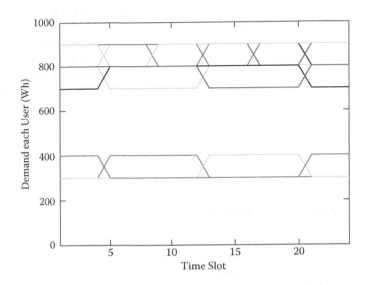

Figure 2.10   Demand of each user with the proposed demand management scheme, over the 24 one-hour time slots.

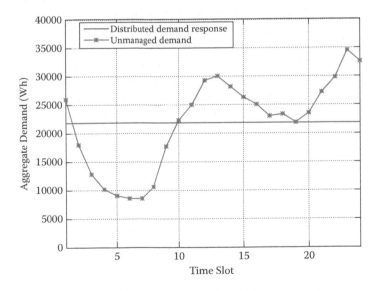

Figure 2.11   Aggregate demand, over all time slots, of the distributed demand management scheme. Unmanaged demand distribution is shown for comparison.

limitation in the power users may request to the generating station, except for the price they are willing to pay.

As we have seen, this DSM model allows us to get quite some insight into the behavior of the users, who accommodate their power consumption to the dynamic pricing policy to minimize both their expenses and the aggregate power demand. Nevertheless, as introduced in Section 2.1, future smart grids are much more complex systems, [7–11] and the DSM model needs to be expanded to account for smart grid features and limitations. In what follows, we will try to explain how DSM may still offer appealing results when considering multiple power-limited energy sources and storage capabilities.

2.5.1 Distributed Generation

Section 2.1 has thoroughly discussed the role of DG in smart grids. Briefly, dispersing energy sources all over the grid brings several benefits, such as accounting for renewable sources and reducing source-load distances and their associated dissipation losses [61]. Moreover, in a properly designed grid, each source may charge with the surrounding loads and (partially) supply their energy needs during main-utility blackouts (islanding) [13].

From the DSM viewpoint, DG may be easily treated by assuming that all energy sources concentrate their energy production in a common pool that serves the loads. However, more degrees of freedom may be added to the problem, resulting in much more interesting systems. Indeed, in a liberalized energy market, users will be given the opportunity to decide which source they draw energy from [61,62]. Observe that much more attention should be paid to the maximum available power. In the previous simplified model, the total available power was assumed to be infinite, while the effective load was kept limited by price-based disincentives. With distributed generators, which are typically small due to economic and environmental reasons [12,61,63], capacity constraints cannot be neglected without jeopardizing model validity.

Unfortunately, technical literature about synergies between DG and DSM does not seem to be very flourishing, and little mention is given to the problem. As an example, A.-H. Mohsenian-Rad et al. claim that, with little effort, DG can be supported by both their centralized [64] and distributed [65] solutions. Such an extension, however, is left for future work.

A different approach is proposed by M. Fahrioglu et al. [66], who understand DG as a sort of alternative to DSM. The paper discusses cases when the DSM principle of users rearranging their load-time profile toward a constant aggregate load is not the best solution. Instead, they propose that sources at customer sites should be activated at high demand occurrences to relieve grid load. Some examples are given where the latter approach results in higher global benefits.

2.5.2 Distributed Storage

Efficient storage of high amounts of energy presents several technical inconveniences that limit storage diffusion in classic electrical grids [12,17]. As discussed in Section 2.1, however, the evolution of the grid to its new smart version has brought new enthusiasm to the analysis and design of storing systems. There are two main reasons behind this interest. First, storage capability would allow customers to buy energy at a low price during off-peak hours and to use it later, when the grid experiences a high load and energy price increases. Second, the intermittent nature of renewable sources requires storage as an intermediary agent to match generation profiles with load profiles. Furthermore, the development of new technologies makes storage a more feasible feature for the smart grid. For instance, it is foreseen that large fleets of electrical vehicles will populate our streets in the near future. During parking/charging hours, their batteries will be available as a storage resource for the grid [19,20].

Dealing with distributed storage (DS) in DSM should not raise difficult problems. As suggested in the aforementioned works by A.-H. Mohsenian-Rad et al. [64,65], discharging storage devices can be modeled as negative loads. Applying this intuitive approach into a game theory DSM framework, P. Vytelingum et al. [67] show that distributed micro-storage (e.g., 4 kWh storage capacity per customer) results in a profitable, efficient electrical grid. In particular, they consider a mixed system where some customers do install storage devices and others do not. Interestingly enough, the highest savings are obtained when only a relatively low fraction of the customers (around 38% in their example) employ storage. This is due to the fact that electricity prices are decided on a DSM strategy and actions aimed at local benefits (i.e., storage) have positive effects on the whole grid.

2.5.3 Issues and Comments

Even though outside the scope of this chapter, it is worth mentioning here that some technical problems need to be faced before a practical implementation of the above DSM techniques could take place in real grids. With no pretension to be exhaustive, we summarize hereafter the salient points and direct interested readers to the bibliography [7,13,61,62,68] and references therein for further details.

The main problem will probably come from the control complexity of the new smart grid. Indeed, the introduction of DG and DS, together with the envisaged real-time supply demand adjustments (as a result of DSM strategies), will add instability sources to an already delicate system such as the power grid. Frequency, phases, and voltage levels will be affected by modifications in the dynamic interconnection of distributed energy sources. Dynamic energy control protocols (DECPs) and just-in-time/just-in-place procedures are being developed to cope with these issues, which are exacerbated by the discontinuous nature of renewable sources and by load variations. Actuators like flexible AC transmission system (FACTS) are also necessary. These devices offer the possibility to compensate reactive power flows in a much faster and precise way than classical solutions like shunt capacitors or generator tuning.

Another issue, untouched by this chapter, is the intrinsic uncertainty of physical systems like electric grids. Even accepting the (already strong) assumption that all system parameters may be measured and reported with no errors, there exist a lot of quantities that cannot be known in advance at planning time. A significant example is the instantaneous power offered by renewable energy plants, which is strongly dependent on, e.g., weather conditions (wind strength, cloud cover, precipitations, and so forth). Centralized DSM approaches may deal with system uncertainties by means of stochastic programming and, more generally, of the theory of optimization under stochastic constraints. Examples in this direction are the works by A. J. Conejo et al. [29] and J.-S. Roy and A. Lenoir [69]. On the other hand, we are not aware of any attempt to include uncertainty in distributed DSM. Bayesian games with incomplete information [53,70] seem to offer a possible framework and are probably worth some investigation.

Communications will also play a fundamental role in future smart grids. [71] All the distributed control procedures described above require a continuous exchange of information among the involved parties. Besides, the DSM strategies discussed in this chapter are based on knowledge of user load profiles at the supplier side and knowledge of price tariffs at the user side. The deployment of the so-called AMI in densely populated areas is a great challenge due to the significant volume of signaling that is generated.

To conclude, we should also comment on a social/behavioral issue (rather than technical) that is quite related with the last point. Power grid customers have long been used to flat, centrally regulated energy rates. It will probably take some time, education efforts, and of course, visible economic incentives to motivate average users to get full advantage of the new dynamic, real-time fluctuating pricing policies.

2.6 Conclusion

Electric grids are undergoing a technological revolution that will change the way electricity is produced, sold, and distributed. In the chapter, we argued that such advances are setting the ground for new demand management schemes. First, advances in communications and automated meter reading will enable timely interactions between consumers and the utility. Second, the incorporation of renewable energy sources, of a distributed nature, and electric vehicles providing storage capabilities will increase the benefits of demand management. The analysis of deregulated wholesale markets reveals that economic factors induce a certain degree of adaptation of energy consumption to available supply. Therefore, additional benefits can be achieved if utilities and energy retailers are able to better control their demand through end user demand management schemes. Two approaches that are gaining momentum were described: direct load control, whereby the utility assumes control of user electricity demand and optimizes it according to a global criterion; and pricing schemes, which provide incentives for end users to manage their demand and reach a global, socially advantageous solution. In this context, game theory provides the adequate set of tools that characterize interactions among users and the pricing signals. In particular, we proposed congestion games, which provide a framework that models the demand management problem and reaches an equilibrium solution resulting in a flat aggregate demand.

Acknowledgments

This work was partially supported by Generalitat de Catalunya under grant 2009-SGR-940, and by the Spanish government under project TEC2010-21100 (SOFOCLES).

References

1. A. Abur and A. Gómez Expósito. *Power system state estimation. Theory and implementation.* (Marcel Dekker, New York, 2004).
2. BLACKOUT2003. Final report on the August 14, 2003, blackout in the United States and Canada: Causes and recommendations. Report, U.S.-Canada Power System Outage Task Force (April 2004). https://reports.energy.gov/BlackoutFinal-Web.pdf.
3. CBC2003. The great blackout of 2003. The CBC Digital Archives website. Canadian Broadcasting Corporation—retrieved March 31, 2011 (August 14, 2003). http://archives.cbc.ca/science_technology/energy_production/clips/13545/.
4. J. Hooper. Italy's blackout raises questions. *The Guardian* (September 30, 2003). http://www.guardian.co.uk/world/2003/sep/30/italy.johnhooper.
5. AP2005. 100 million Indonesians affected by power outage. *AP Worldstream* (August 18, 2005).
6. AFP2009. Brazil insists short circuits, bad weather caused blackout. Agence France-Press (AFP) (November 16, 2009).
7. FCC2010. Connecting America: The national broadband plan. Report, Federal Communications Commission (2010). http://www.broadband.gov/download-plan/.
8. ETPSG2005. Towards smart power networks. Report, Directorate-General for Research Sustainable Energy Systems. (2005). http://www.smartgrids.eu/.

9. ETPSG2006. European Technology Platform SmartGrids: Vision and strategy for Europe's electricity networks of the future. Report, European Commission, Directorate-General for Research Sustainable Energy Systems (2006). URL http://www.smartgrids.eu/.

10. ETPSG2007. European Technology Platform SmartGrids: Strategic research agenda for Europe's electricity networks of the future. Report, European Commission, Directorate-General for Research Cooperation Energy (2007). URL http://www.smartgrids.eu/.

11. ETPSG2010. European Technology Platform SmartGrids: Strategic deployment document for Europe's networks of the future. Report, European Commission, Directorate-General for Research Energy (2010). URL http://www.smartgrids.eu/.

12. G. L. Johnson, S. Rahman, and R. A. Messenger. Electric power generation: Non-conventional methods. In ed. L. L. Grigsby, *The electric power engineering handbook*, chap. 1. CRC Press, Boca Raton, FL, (2000).

13. A. Vojdani. Smart integration. *IEEE Power Energy Mag.* **6**(6), 71–79 (2008).

14. IEEE. IEEE P1547.4 draft guide for design, operation, and integration of distributed resource island systems with electric power systems. Draft, Institute of Electrical and Electronics Engineers (2011). http://grouper.ieee.org/groups/scc21/1547.4/1547.4_index.html.

15. J. D. Kueck, R. H. Staunton, S. D. Labinov, and B. J. Kirby. Microgrid energy management system (2003). http://www.ornl.gov/sci/btc/apps/Restructuring/TM2002-242.pdf.

16. FERC. Grid 2030 a national vision for electricity's second 100 years. Report, U.S. Department of Energy, Office of Electric Transmission and Distribution (2003). http://www.ferc.gov/eventcalendar/files/20050608125055-grid-2030.pdf.

17. B. Roberts. Capturing grid power. *IEEE Power Energy Mag.* **7**(4), 32–41 (2009).

18. R. Walawalkar and J. Apt. Market analysis of emerging electric energy storage systems. Technical report, National Energy Technology Laboratory (2008). http://www.netl.doe.gov/energy-analyses/pubs/Final%20Report-Market%20Analysis%20of%20Emerging%20Electric%20Energy%20Sto.pdf.

19. W. Kempton and J. Tomić. Vehicle-to-grid power fundamentals: Calculating capacity and net revenue. *Journal of Power Sources* **144**(1), 268–279 (2005).

20. W. Kempton and J. Tomić. Vehicle-to-grid power implementation: From stabilizing the grid to supporting large-scale renewable energy. *Journal of Power Sources* **144** (1), 280–294 (2005).

21. EEP2007. European electricity projects 2002–2006. Report, European Commission, Directorate-General for Research Sustainable Energy Systems (2007). http://ec.europa.eu/research/energy/.

22. D. Wight et al. Assessment of demand response and advanced metering. Staff report, FERC Federal Energy Regulatory Commission, US (2011). http://www.ferc.gov/legal/staff-reports/.

23. The Brattle Group and Freeman, Sullivan & Co. and Global Energy Partners, LLC. A national assessment of demand response potential. Staff report (June 2009). http://www.ferc.gov/legal/staff-reports/.

24. D. Wight et al. Assessment of demand response and advanced metering. Staff report FERC Federal Energy Regulatory Commission, US (2007). http://www.ferc.gov/legal/staff-reports/.

25. S. Kärkkäinen et al. Integration of demand side management, distributed generation, renewable energy sources and energy storages. State of the art report 1, International Energy Agency, Demand Side Management Program (December 2008).

26. S. Kärkkäinen et al. Integration of demand side management, distributed generation, renewable energy sources and energy storages. Vol. 2: Annexes. State of the art report 1, International Energy Agency, Demand Side Management Program (December 2008).

27. REE. Red Eléctrica de España. Power demand tracking in real time (on-line tool). http://www.ree.es/ingles/operacion/curvas_demanda.asp.

28. P. Stephenson and M. Paun. Electricity market trading. *Power Engineering Journal.* **15** (6), 277–288 (2001).

29. A. J. Conejo, M. Carrión, and J. M. Morales. *Decision making under uncertainty in electricity markets.* (Springer US, Boston, 2010).

30. CNE. Comisión Nacional de la Energía. Spanish regulator electricity market. http://www.eng.cne.es.

31. A. Molina, A. Gabaldon, J. A. Fuentes, and C. Alvarez. Implementation and assessment of physically based electrical load models: Application to direct load control residential programs. *IEE Proceedings Generation, Transmission & Distribution* **150**(1), 61–66 (2003).

32. D. Bargiotas and J. D. Birdwell. Residential air conditioner dynamic model for direct load control. *IEE Proceedings Generation, Transmission & Distribution.* **3**(4), 2119–2126 (1998).

33. G. G. Virk and D. L. Loveday. Model-based control of HVAC applications. In *Proceedings of the IEEE Conference on Control Applications*, vol. 3, pp. 1861–1866 (1994).

34. K. Tomiyama, J. P. Daniel, and S. Ihara. Modeling air conditioner load for power system studies. *IEEE Transactions Power Systems* **13**(2), 414–420 (1998).

35. C. F. Walker and I. L. Pokoski. Residential load shape modeling based on customer behavior. *IEEE Transactions on Power Apparatus Systems* **104**(7), 1703–1711 (1985).

36. S. Ihara and C. Schweppe. Physically based modeling of cold load pickup, *IEEE Transactions on Power Apparatus Systems* **100**, 4142–42150 (1981).

37. GAD. GAD project: Active and efficient electric consumption management (Gestión Activa de la Demanda, Spanish project funded by CENIT program) (2007–2010). http://www.proyectogad.com.

38. H. Lee and C. Wilkins. A practical approach to appliance load control analysis: A water heater case study. *IEEE Transactions on Power Apparatus Systems* **102**(4) (1983).

39. A. Cohen and C. Wang. An optimization method for load management scheduling. *IEEE Transactions Power Systems* **3**(4), 612–618 (1988).

40. Z. Popovic. Determination of optimal direct load control strategy using linear programming. In *CIRED* (1999).

41. C. Kurucz. A linear programming model for reducing system peak through customer load control programs. *IEEE Transactions Power Systems* **11** (4), 1817–1824 (1996).

42. K.-H. Ng and G. B. Sheble. Direct load control—A profit-based load management using linear programming. *IEEE Transactions Power Systems* **13**(2), 688–694 (1998).

43. D. C. Wei and N. Chen. Air conditioner direct load control by multipass dynamic programming. *IEEE Transactions Power Systems* **10**(1), 307–313 (1995).

44. K.-Y. Huang and Y.-C. Huang. Integrating direct load control with interruptible load management to provide instantaneous reserves for ancillary services. *IEEE Transactions Power Systems* **19**(3), 1626–1634 (2004).

45. H.-T. Yang and K.-Y. Huang. Direct load control using fuzzy dynamic programming. *IEE Proceedings Generation, Transmission & Distribution* **146**(3), 294–300 (1999).

46. C.-M. Chu and T.-L. Jong. A novel direct air-conditioning load control method. *IEEE Transactions Power Systems* **23**(3), 1356–1363 (2008).

47. L. Yao, W.-C. Chang, and R.-L. Yen. An iterative deepening genetic algorithm for scheduling of direct load control. *IEEE Transactions Power Systems* **20**(3), 1414–1421 (2005).

48. I. Cobelo. Active control of distribution networks. PhD thesis, School of Electrical and Electronic Engineering, Faculty of Engineering and Physical Sciences, University of Manchester (2005).

49. N. Ruiz, I. Cobelo, and J. Oyarzabal. A direct load control model for virtual power plant management. *IEEE Transactions Power Systems* **24**(2), 959–966 (2009).

50. J. Walrand. *Economic models of communication networks* (Springer US, Boston, 2008).

51. J. G. Wardrop. Some theoretical aspects of road traffic research. *Proceedings of the Institution of Civil Engineers, Part II* **1**(36), 325–378 (1952).

52. R. W. Rosenthal. A class of games possessing pure-strategy Nash equilibria. *International Journal of Game Theory* **2**(1), 65–67 (1973).

53. D. Fudenberg and J. Tirole. *Game theory* (MIT Press, Cambridge, MA, 1991).

54. A. Salehian. Dynamic game theory model for the power transmission grid. In *Proceedings of the IEEE PES Power Systems Conference and Exposition*, vol. 3, pp. 1449–1452 (October 10–13, 2004).

55. A.-H. Mohsenian-Rad, V. W. Wong, J. Jatskevich, and R. Schober. Optimal and autonomous incentive-based energy consumption scheduling algorithm for smart grid. In *Proceedings of the IEEE PES Conference on Innovative Smart Grid Technologies*, pp. 1–6, Gaithersburg, MD (January 19–21, 2010).

56. C. Ibars, M. Navarro, and L. Giupponi. Distributed demand management in smart grid with a congestion game. In *Proceedings of IEEE First International Conference on Smart Grid Communications, IEEE SmartGridComm 2010*, Gaithersburg, MD (October 4–6, 2010).

57. D. A. Monderer and L. S. Shapley. Potential games. *Games and Economic Behavior* **14**(1), 124–143 (1996).

58. B. A. Carreras, V. E. Lynch, I. Dobson, and D. E. Newman. Critical points and transitions in an electric power transmission model for cascading failure blackouts. *Chaos* **12**(4), 985–994 (2002).
59. A. Fabrikant, C. Papadimitriou, and K. Talwar. The complexity of pure Nash equilibria. In *Proceedings of the 36th Annual ACM Symposium on Theory of Computing (STOC)*, pp. 604–612 (2004).
60. D. P. Bertsekas. *Network optimization* (Athena Scientific, 1998).
61. A. Keyhani, M. N. Marwali, and M. Dai. *Integration of green and renewable energy in electric power systems* (John Wiley & Sons, New York, 2010).
62. M. D. Ilic. From hierarchical to open access electric power systems. *Proceedings IEEE* **95**(5), 1060–1084 (2007).
63. R. C. Armstrong and E. J. Moniz. Report of the energy council. Report, MIT Energy Research Council, Cambridge, MA (2006). http://web.mit.edu/mitei/about/erc-report-final.pdf.
64. A.-H. Mohsenian-Rad and A. Leon-Garcia. Optimal residential load control with price prediction in real-time electricity pricing environments. *IEEE Transactions Smart Grid*. **1** (2), 120–133 (2010).
65. A.-H. Mohsenian-Rad, V. W. S. Wong, J. Jatskevich, R. Schober, and A. Leon-Garcia. Autonomous demand-side management based on game-theoretic energy consumption scheduling for the future smart grid. *IEEE Trans. Smart Grid*. **1** (3), 320–331 (2010).
66. M. Fahrioglu, R. H. Lasseter, F. L. Alvarado, and T. Yong. Integrating distributed generation technology into demand management schemes. In *Proceedings of the IEEE PES/IAS Conference on Sustainable Alternative Energy*, Valencia, Spain (September 28–30, 2009).
67. P. Vytelingum, T. D. Voice, S. D. Ramchurn, A. Rogers, and N. R. Jennings. Agent-based micro-storage management for the smart grid. In *Proceedings of the 9th International Conference on Autonomous Agents and Multiagent Systems (AAMAS 2010)*, Toronto, Canada (May 10–14, 2010).
68. P. Ross, J. Romero, W. Jones, A. Bleicher, J. Calamia, J. Middleton, R. Stevenson, S. Moore, S. Upson, D. Schneider, E. Guizzo, P. Fairley, T. Perry, and G. Zorpette. Top 11 technologies of the decade. *IEEE Spectr.* **48**(1 (INT)), 23–59 (2011).
69. J.-S. Roy and A. Lenoir. Nonparametric approximation of non-anticipativity constraints in scenario-based multistage stochastic programming. *Kybernetika* **44**(2), 171–184 (2008).
70. Y. Shoham and K. Leyton-Brown. *Multiagent systems. Algorithmic, game-theoretic, and logical foundations* (Cambridge University Press, 2009). http://www.masfoundations.org/.
71. GCENERGY2010. Communications requirements of smart grid technologies. Report, U.S. Department of Energy (2010). http://www.gc.energy.gov/documents/Smart_Grid_Communications_Requirements_Report_10-05-2010.pdf.

Chapter 3

Efficient Management of Locally Generated Powers in Microgrids

Tomaso Erseghe, Stefano Tomasin, and Paolo Tenti

Contents

A residential microgrid is a subsection of the grid associated to a small village or a neighborhood that includes distributed energy sources. In order to sustain the loads of the microgrid, power can be provided either by the utility or by the distributed sources at costs that depend on the energy market. In this chapter we focus on optimizing the energy sources in order to minimize the cost incurred by the microgrid, to make it smarter. In our model, costs are proportional to the injected power. Transmission line losses, which may significantly contribute to the waste of energy and cost rise, are also taken into account in the optimization process. We investigate both the optimal solution to the problem, which is computationally expensive, and a suboptimal solution based on the steepest descent algorithm.

3.1 Smart Microgrids

The spread of distributed energy sources (DESs), such as photovoltaic panels, wind generators, and micro turbines, has changed the role of the customers of the main electrical utility. They are also producers that can actually not only sustain their own loads but also sell energy to the grid. As a result, the new character of *prosumer* (producer and consumer; term first used in this context in [1]) emerges as a distinctive feature of current and future grids. Today, each prosumer interacts directly with the utility and can adjust his own usage and production according to energy market prices and agreements with the utility. However, as the number of DESs increases, we envision a scenario where prosumers living in the same area (e.g., a small village or a town neighborhood) may join forces and operate in synergy, with the objective of optimizing the production and bargain power with the utility. The aggregation of prosumers brings into the scene a new entity, at an intermediate level between the utility and the individuals, i.e., the *microgrid*.

Depending on agreements with the utility, the microgrid may find it useful to first satisfy the internal energy demand using DESs, and then sell the remaining energy to the grid. In this case sharing both DESs and loads in a larger community may provide a way of averaging requests and offers. Another benefit of sharing DESs comes from the possibility of operating in an insulated manner, where the microgrid is disconnected from the grid whenever the cost of energy from the utility becomes excessive, or when outage occurs on the grid. Actually, this scenario may be favorable even to the utility that—knowing that the neighborhood can operate independently—may disconnect it in order to avoid outage, and may concentrate the energy delivery to subgrids that have no alternative sources.

There are various issues arising in a smart management of the microgrid, which may bring significant benefits to users. First, the energy provided by DESs must be optimized according to their costs and the cost of the energy provided by the utility. A second issue pertains to line losses, since distances between loads and sources may become relevant, and losses may increase the costs of energy distribution. Other issues are related to the instantaneous current distribution for balancing rapid changes of loads or sources, and to the use of DESs to compensate reactive powers in the microgrid. All these issues have effects on the network on different timescales. Cost optimization can be performed on the timescale of hours, compatible with the possibility of switching DESs on and off. Compensation of reactive power and current balancing to face fluctuations of DES supply should instead be operated on a reduced timescale (less than a second) in order to be effective.

Given the above scenario, the purpose of this chapter is threefold: we aim at proposing a model that takes into account the said issues, we aim at providing a bound for the potential benefits of an optimal microgrid management, and we aim at investigating a possible implementation of the optimization process. Specifically, by assuming a linear description of the electrical network, we propose to optimize the currents injected by each DES with the aim of minimizing the total cost under a constraint on the maximum power that can be provided by both DESs and the utility. We also request DESs to limit reactive current flow, by imposing a bound on the quality factor at the interface with the utility. We finally assume a perfect knowledge of the microgrid parameters (line impedances, load description, and costs), which are also assumed to be time invariant. This problem turns out to be a nonconvex quadratic constrained quadratic problem (QCQP), whose solution requires sophisticated tools. In order to explore a practical solution, we investigate a suboptimal algorithm based on a constrained steepest descent approach.

3.2 Related Literature

Microgrids are new entities, whose optimization is a subject of study only in the recent years. However, some existing literature has dealt with the optimization of the network to improve its efficiency. For example, in [2] the reduction of losses in power transmission is targeted by suitably injecting currents (also termed switching), and the problem is formulated as a linear programming problem, which is solved by a two-step approximation: the first step updates the current and the second step updates the load flow, leading to the optimal solution. The minimization of overloads has been pursued in [3], where an algorithm is proposed to optimize the switching for relieving overloads and voltage violations. The optimization has been addressed from an economic point of view in [4], considering that the prosumer can bid capacity spot auctions to derive congestion revenues on various timescales (daily, monthly, or multimonthly auctions). Another part of the literature has been focused on security issues related to the network optimization, as initiated by [5][6]. More recently, optimization of electric power has been considered jointly with the optimization of other energy sources (including heat and power) in [7].

3.3 MicroGrid Model

We first provide a model of the smart microgrid, from both an electrical and an economic point of view. The model will include the main characteristics of the microgrid, i.e., a description of sources (with their costs), distribution lines, and loads.

3.3.1 Electrical Model

We consider a microgrid with N_{CC} interfaces with the utility grid, denoted point of common couplings (PCCs). Within the electrical network of the microgrid N_{EI} nodes are power electronic interfaces (PEIs), while the remaining N_L nodes are loads, for a total of $N = N_{CC} + N_{EI} + N_L$ nodes, each identified by an index in the range 1 to N. We collect indices of PCCs, PEIs, and loads into the sets $\mathcal{I}_{CC}$, $\mathcal{I}_{EI}$, and $\mathcal{I}_L$, respectively. Assume K active harmonics, and no Direct Current (DC), and denote quantities referring to the kth harmonic with the superscript k, $k = 1, \dots, K$. The three types of nodes are modeled as follows:

- PCCs are modeled as ideal voltage generators with open-circuit voltage $V_{0,n}^{(k)}$, $n \in \mathcal{I}_{CC}$.
- PEIs are modeled as ideal current generators with closed-circuit current $I_{0,n}^{(k)}$, $n \in \mathcal{I}_{EI}$.
- Loads are modeled as impedances $Z_n^{(k)}$, $n \in \mathcal{I}_L$.

In all nodes $I_n^{(k)}$ and $V_n^{(k)}$ represent, respectively, the output node current and the node voltage. We thus have the Kirchhoff equations at each node:

$$V_n^{(k)} = V_{0,n}^{(k)}, \quad n \in \mathcal{I}_{CC}$$
$$I_n^{(k)} = I_{0,n}^{(k)}, \quad n \in \mathcal{I}_{EI} \tag{3.1}$$
$$I_n^{(k)} = -\frac{V_n^{(k)}}{Z_n^{(k)}}, \quad n \in \mathcal{I}_L$$

The line impedance associated with connection between node n and node m is denoted with $Z_{\text{line } n,m}^{(k)}$. An example of a microgrid is shown in Figure 3.1.

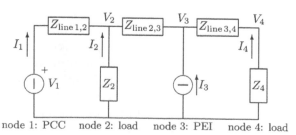

node 1: PCC   node 2: load   node 3: PEI   node 4: load

We assume that the value of all impedances is known at the control unit. Moreover, PCC open-circuit voltages $V_{0,n}^{(k)}$ are known. Instead, PEI currents $I_{0,n}^{(k)}$ are to be set according to some target optimization strategy. This requires that a full understanding of network voltages and currents is available as a function of PEI currents. By assuming a linear network model, for the kth harmonic this relation has the vector and matrix form[1]

$$
\begin{aligned}
I_n^{(k)} &= \left(a_n^{(k)}\right)^* I_{0,\mathrm{EI}}^{(k)} + \alpha_n^{(k)}, \quad n \in \mathcal{I}_{\mathrm{CC}} \\
V_n^{(k)} &= \left(b_n^{(k)}\right)^* I_{0,\mathrm{EI}}^{(k)} + \beta_n^{(k)}, \quad n \in \mathcal{I}_{\mathrm{EI}}
\end{aligned}
\tag{3.2}
$$

where

- The column vector $I_{0,\mathrm{EI}}^{(k)} = [I_{0,n}^{(k)}]_{n \in \mathcal{I}_{\mathrm{EI}}}$ carries PEI input currents.
- Vectors $a_n^{(k)}$ and $b_n^{(k)}$ depend on the network structure (i.e., line and node impedances).
- Scalars $\alpha_n^{(k)}$ and $\beta_n^{(k)}$ also depend on PCC open-circuit voltages.

The explicit expression (3.2) as a function of network parameters is provided in Appendix A.

We define the active electrical power generated or consumed by node n as[2]

$$
P_n = \sum_{k=1}^{K} 2\,\Re\left[\left(V_n^{(k)}\right)^* I_n^{(k)}\right]
\tag{3.3}
$$

Generation capabilities are limited by a maximum active power U_n. PCC and PEI can also in general absorb active power (thus behaving as controlled loads), and the maximum active power that can be absorbed by node n is denoted L_n. Hence we have

$$
L_n \leq P_n \leq U_n, \quad n \in \{\mathcal{I}_{\mathrm{CC}} \cup \mathcal{I}_{\mathrm{EI}}\}
\tag{3.4}
$$

In order to limit the reactive power exchanged with the PCCs, we impose a quality factor constraint. In particular, we define the power factor at PCCs as

$$
\lambda_n = \frac{|P_n|^2}{\displaystyle\sum_{k=1}^{K} 2\left|V_n^{(k)}\right|^2 \sum_{k=1}^{K} 2\left|I_n^{(k)}\right|^2}, \quad n \in \mathcal{I}_{\mathrm{CC}}
\tag{3.5}
$$

---

[1] x^* is the transpose and complex conjugate of complex vector x.
[2] $\Re[x]$ and $\Im[x]$ represent the real and imaginary parts of x, respectively.

and we impose

$$\lambda_n \geq 1 - \epsilon$$

with ϵ a small number.

3.3.2 Economic Model

The costs associated with power sources can be classified as fixed (e.g., amortization costs, maintenance costs, etc.) or variable (e.g., fuel costs for micro turbines). Variable costs may be subject to various models and they are in general related to the generated active power.

In the following we consider a simple model where the cost associated to each generator is proportional to the active power by a proper proportionality cost factor c_n, $n \in \{\mathcal{I}_{EI} \cup \mathcal{I}_{CC}\}$, depending on the type of source. We are thus ignoring fixed costs. Although this model is simple, it already provides an important insight into the dynamics of the network and the costs. Moreover, it can be considered a first-order approximation of a more elaborate cost function, which can be applied when powers are within a limited range.

The total variable costs of the microgrid can then be written as

$$c = \sum_{n \in \mathcal{I}_{CC} \cup \mathcal{I}_{EI}} c_n P_n \tag{3.6}$$

We further assume that renewable sources have zero variable costs, i.e., $c_n = 0$, with n being the index of a PEI powered by a renewable source.

3.4 MicroGrids Get Smart

Given the electrical and economic model, our objective is to minimize the operation cost of the microgrid, while supporting loads and satisfying the quality constraints. In formulas we have

$$\text{minimize } c = \sum_{n \in \mathcal{I}_{CC} \cup \mathcal{I}_{EI}} c_n P_n$$

$$\text{with respect to } I_n^{(k)}, \, n \in \mathcal{I}_{EI}, \, k = 1, \ldots, K$$

$$\text{subject to } L_n \leq P_n \leq U_n, \, n \in \mathcal{I}_{CC} \cup \mathcal{I}_{EI}$$

$$\lambda_n \geq 1 - \epsilon, \, n \in \mathcal{I}_{CC}$$

(3.7)

We first rewrite (3.7) as a function of PEI currents collected in vector $I = [I_{0,EI}^{(k)}]_{k=1,...,K}$ of length $K N_{EI}$. Then we show that it is a nonconvex QCQP problem and investigate methods for its solution.

Power constraints: By use of (3.2), the power (3.3) can be written in the complex valued quadratic form

$$P_n = I^* R_n I + 2\Re[I^* r_n] + \rho_n \qquad (3.8)$$

where matrix R_n, vector r_n and scalar ρ_n are specified in the following. For PCC we have[3]

$$n \in \mathcal{I}_{CC} : \begin{cases} R_n = 0 \\ r_n = \displaystyle\sum_{k=1}^{K} e_{k,K} \otimes \left(V_{0,n}^{(k)} a_n^{(k)} \right) \\ \rho_n = \displaystyle\sum_{k=1}^{K} 2\Re\left[V_{0,n}^{(k)} \left(\alpha_n^{(k)} \right)^* \right] \end{cases} \qquad (3.9)$$

with $e_{k;K}$ a column vector of length K having a unique nonzero entry of value 1 at position k. Instead, for PEI it is

$$n \in \mathcal{I}_{EI} : \begin{cases} R_n = \displaystyle\sum_{k=1}^{K} \mathrm{diag}(e_{k,K}) \otimes \left(b_n^{(k)} e_{\varphi(n),N_{EI}}^* + e_{\varphi(n),N_{EI}} \left(b_n^{(k)} \right)^* \right) \\ r_n = \displaystyle\sum_{k=1}^{K} e_{k,K} \otimes \left(\beta_n^{(k)} e_{\varphi(n),N_{EI}} \right) \\ \rho_n = 0 \end{cases}$$

$$(3.10)$$

where $\varphi(n)$ denotes the position of $I_{0,n}^{(k)}$, $n \in \mathcal{I}_{EI}$, in vector $I_{0,EI}^{(k)}$. Note that, by construction, R_n are Hermitian matrices.
A real valued form can also be obtained by use of the functionals $\dot{}$ and $\ddot{}$, respectively,

$$\dot{v} = \begin{bmatrix} \Re[v] \\ \Im[v] \end{bmatrix}, \qquad \ddot{M} = \begin{bmatrix} \Re[M] & -\Im[M] \\ \Im[M] & \Re[M] \end{bmatrix} \qquad (3.11)$$

to have

$$P_n = \dot{i}^T \ddot{R}_n \dot{i} + 2 \dot{i}^T \dot{r}_n + \rho_n \qquad (3.12)$$

---

[3] $x \otimes y$ is the Kronecker product between x and y.

Cost function: From (3.6) and (3.8) the cost function can be written in the complex quadratic form

$$c = I^* R I + 2 \Re[I^* r] + \rho \tag{3.13}$$

with

$$R = \sum_{n \in \mathcal{I}_{CC} \cup \mathcal{I}_{EI}} c_n R_n , \qquad r = \sum_{n \in \mathcal{I}_{CC} \cup \mathcal{I}_{EI}} c_n r_n , \qquad \rho = \sum_{n \in \mathcal{I}_{CC} \cup \mathcal{I}_{EI}} c_n \rho_n \tag{3.14}$$

By mimicking (3.12), we also have the real valued quadratic form

$$c = \dot{\imath}^T \ddot{R} \dot{\imath} + 2 \dot{\imath}^T \dot{r} + \rho \tag{3.15}$$

Power quality constraint: Starting from (3.2) the denominator of (3.5) can be written as

$$I^* D_n I + 2 \Re[I^* d_n] + \delta_n \tag{3.16}$$

where

$$n \in \mathcal{I}_{CC} : \begin{cases} D_n = K_n \sum_{k=1}^{K} \mathrm{diag}(e_{k,K}) \otimes \left(a_n^{(k)} \left(a_n^{(k)} \right)^* \right) \\[2mm] d_n = K_n \sum_{k=1}^{K} e_{k,K} \otimes \left(\alpha_n^{(k)} a_n^{(k)} \right) \\[2mm] \delta_n = K_n \sum_{k=1}^{K} \left| \alpha_n^{(k)} \right|^2 , \quad K_n = 4 \sum_{k=1}^{K} \left| V_{0,n}^{(k)} \right|^2 \end{cases} \tag{3.17}$$

Since $R_n = 0$ when $n \in \mathcal{I}_{CC}$, using (3.8) the numerator of (3.5) can be written as

$$\left| I^* r_n + r_n^* I + \rho_n \right|^2 \tag{3.18}$$

Note that (3.18) cannot be written in the standard complex quadratic form (3.16). A quadratic form can instead be obtained by use of real valued quantities, since $P_n = 2 \dot{\imath}^T \dot{r}_n + \rho_n$, $n \in \mathcal{I}_{CC}$. By thus defining

$$n \in \mathcal{I}_{CC} : \begin{cases} N_n = 4 \dot{r}_n \dot{r}_n^T \\[1mm] n_n = 2 \rho_n \dot{r}_n \\[1mm] v_n = \rho_n^2 \end{cases} \tag{3.19}$$

where we used (3.11), we have

$$\lambda_n = \frac{\mathbf{i}^T \mathbf{N}_n \mathbf{i} + 2\, \mathbf{i}^T \mathbf{n}_n + v_n}{\mathbf{i}^T \mathbf{D}_n \mathbf{i} + 2\, \mathbf{i}^T \mathbf{d}_n + \delta_n} \tag{3.20}$$

The power quality constraint $\lambda_n \geq 1 - \epsilon$ in (3.7) can thus be expressed as

$$\mathbf{i}^T[(1-\epsilon)\ddot{\mathbf{D}}_n - \mathbf{N}_n]\mathbf{i} + 2\, \mathbf{i}^T[(1-\epsilon)\dot{\mathbf{d}}_n - \mathbf{n}_n]$$

$$\tag{3.21}$$

$$+ [(1-\epsilon)\delta_n - v_n] \leq 0$$

for $n \in \mathcal{I}_{\mathrm{CC}}$.

3.4.1 Target Optimization Revisited

In order to collect results together in compact form, we define the total number of constraints as $N_{\mathrm{tot}} = 3N_{\mathrm{CC}} + 2N_{\mathrm{EI}}$, and define matrices

$$C_n = \begin{bmatrix} \ddot{\mathbf{R}}_n & \mathbf{p}_n \\ \mathbf{p}_n^T & q_n \end{bmatrix}$$

$$\tag{3.22}$$

$$E_n = \begin{bmatrix} (1-\epsilon)\ddot{\mathbf{D}}_n - \mathbf{N}_n & (1-\epsilon)\dot{\mathbf{d}}_n - \mathbf{n}_n \\ (1-\epsilon)\dot{\mathbf{d}}_n^T - \mathbf{n}_n^T & (1-\epsilon)\delta_n - v_n \end{bmatrix}$$

Note that all matrices of (3.22) are real valued and symmetric by construction. We finally introduce the cost matrix

$$C = \begin{bmatrix} \ddot{\mathbf{R}} & \mathbf{p} \\ \mathbf{p}^T & q \end{bmatrix} \tag{3.23}$$

Then, the target function (3.7) can be rewritten in compact quadratic form as

minimize $\mathbf{x}^T C \mathbf{x}$

with respect to $\mathbf{I}$

subject to $\mathbf{x} = \begin{bmatrix} \mathbf{i} \\ 1 \end{bmatrix}$ $\tag{3.24}$

$$L_n \leq \mathbf{x}^T C_n \mathbf{x} \leq U_n, \; n \in \mathcal{I}_{\mathrm{CC}} \cup \mathcal{I}_{\mathrm{EI}}$$

$$\mathbf{x}^T E_n \mathbf{x} \leq 0, \; n \in \mathcal{I}_{\mathrm{CC}}$$

3.4.2 Nonconvexity Statement

Some of the matrices involved in the quadratic form are not positive semidefinite; that is, the cost function and the constraints are in general nonconvex. Although N_n and D_n (and so $\ddot{D}_n$) are positive semidefinite by construction, (3.21) is not.

Furthermore matrices R_n, and so R, are not positive semidefinite. We prove this latter statement by exploiting the block form of R_n, $n \in \mathcal{I}_{EI}$. Specifically, we investigate the structure of the eigenvalue of the Hermitian block $b_n^{(k)} e_{m,N_{EI}}^* + e_{m,N_{EI}} (b_n^{(k)})^*$, where $m = \varphi(n)$. This block identifies the two eigenvalues (the remaining eigenvalues are simply 0 by construction)[4]

$$\lambda_\pm = \Re\left[\left[b_n^{(k)}\right]_m\right] \pm \sqrt{\left\|b_n^{(k)}\right\|^2 - \Im\left[\left[b_n^{(k)}\right]_m\right]^2} \tag{3.25}$$

and the two eigenvectors

$$v_\pm = b_n^{(k)} + \left(\lambda_\pm - \left[b_n^{(k)}\right]_m\right) e_{m,N_{EI}}$$

with

$$\left\|v_\pm\right\|^2 = \left\|b_n^{(k)}\right\|^2 + |\lambda_\pm|^2 - \left|\left[b_n^{(k)}\right]_m\right|^2$$

We have $\Re[[b_n^{(k)}]_m]^2 \le \|b_n^{(k)}\|^2 - \Im[[b_n^{(k)}]_m]^2$, where equality holds when $[b_n^{(k)}]_m$ is the only nonzero element of $b_n^{(k)}$. Then, one of the two eigenvalues in (3.25) is positive and the other one is negative. Hence R_n has K positive eigenvalues, K negative eigenvalues, and the rest 0. Finally note that, by construction, the eigenvalues of $\ddot{R}_n$ are the same as R_n with twice their multiplicity.

3.5 Suboptimal Solution

Solving nonconvex QCQP problems is a tough task, as they belong to the NP-hard class. In order to obtain a practical solution we consider a steepest descent (SD) technique. First note that the optimization problem is subject to various nonconvex constraints, i.e., power and quality constraints. Finding a feasible solution is also a difficult task. Moreover, since the problem is nonconvex, it has in general many local minima and the SD algorithm will converge to one of these local minima.

The approach we follow for the solution is reported in Figure 3.1. All steps of the process are implemented using an SD algorithm, namely:

---

[4] The norm of a vector x is denoted as $\|x\|$. The nth entry of vector x is denoted as $[x]_n$.

(1) Find point I_1 satisfying power constraints only, starting from a random point I_0.
(2) Starting from I_1, find a solution I_2 to the QCQP problem (3.24) without taking into account the quality factor constraint.
(3) Starting from I_2 find a feasible solution I_3 satisfying all constraints.
(4) Further optimize I_3 taking into account all constraints.

Figure 3.1   Suboptimal solution of QCQP problem.

Step 1 looks for a feasible solution satisfying all power constraints. The initial point of this step may be anyone, e.g., $I_0 = 0$, or we can assume that DESs are at nominal voltage and generate their maximum power so injected currents can be computed in a closed form. If no solution can be found we have to randomly search for another starting point until we find a feasible solution.

Step 2 starts from the feasible solution I_1 and aims at minimizing the costs. In this case we are sure to find at least one solution, i.e., I_1.

Step 3 adds the quality factor constraint. As for step 1, also in this case we may not be able to find a feasible solution starting from I_1, and if this occurs, we have to go back to step 1 and look for another starting point I_0.

Step 4 provides a last optimization starting from the fully feasible point found at the previous step.

3.5.1 Steepest Descent Algorithm

All steps of the proposed suboptimal solution are implemented using a SD algorithm, which is detailed in this section. The idea of the SD is to move from an initial starting point along the inverse gradient of the function to be minimized. In the presence of constraints, the movement is conditioned upon the satisfaction of the constraints.

Constrained minimization: Consider the following general constrained minimization problem on the vector variable x

$$\text{minimize } f(x)$$

$$\text{subject to } g_i(x) \le 0, \ i = 1, \dots, Q$$

(3.26)

with x a vector of length M. Let x_0 be an initial feasible solution. Let also $\nabla f(x) : |\mathrm{R}^M \to |\mathrm{R}^M$ be the gradient of $f(x)$ computed at

(a) Start from feasible x
(b) For $m = 1 : M$
(c)   Set $\gamma = \gamma_0$
(d)   While $\gamma \geq \gamma_{min}$
(e)   Set $y = x$, and $d = \nabla f(x) / \|\nabla f(x)\|$
(f)   Set $[y]_m = [x]_m - \gamma [d]_m$
(g)   If $f(y) < f(x)$ and $g_i(y) \leq 0$, $i = 1, \ldots, Q$
(h)   Update $x = y$
(i)   Exit while cycle
(j)   Else
(k)   Set $\gamma = \gamma \cdot \delta_\gamma$

Figure 3.2   SD algorithm update based on backtracking line search.

point x. Then, starting from the feasible point x_0, the SD algorithm updates x using a backtracking line search, as reported in Figure 3.2, where γ_0, γ_{min}, and $\delta_\gamma < 1$ are backtracking (positive) constants.

The SD algorithm moves along the direction of the gradient and the size of the step is determined by γ. In order to ease convergence we first normalize the gradient to its norm and then we try to move along each individual direction of the gradient. This proved to be more effective than moving along the gradient's direction. If the solution is not feasible, or does not provide a reduction of the objective function, the value of γ is reduced.

Feasible solution search: We also apply the SD algorithm to find feasible solutions. This is performed in an incremental manner, starting from a point x satisfying $Q - 1$ constraints, and looking for a point x also satisfying an additional constraint. Specifically, we aim at solving

$$\text{find } x \text{ such that } g_Q(x) \leq 0$$

$$\text{(3.27)}$$

$$\text{subject to } g_i(x) \leq 0, \ i = 1, \ldots, Q - 1$$

Also in this case we use the SD algorithm update of Figure 3.2, which will be stopped as soon as the constraint on $g_Q(x)$ is satisfied.

We now specialize the SD algorithm to each of the steps of Figure 3.1. For all steps the minimization variable is the real valued vector $x = \dot{I}$.

Step (1): The algorithm (3.27) is applied $2N_{CC} + 2N_{EI}$ times in cascade, in order to introduce one power constraint at a time. From (3.4) and

(3.12), let power constraint functions be defined as

$$
g_q(\boldsymbol{x}) = \begin{cases} \boldsymbol{x}^T \ddot{\boldsymbol{R}}_n \boldsymbol{x} + 2\,\boldsymbol{x}^T \dot{\boldsymbol{r}}_n + \rho_n - U_n\,, & q \text{ even, } n \in \mathcal{I}_{CC} \cup \mathcal{I}_{EI} \\ L_n - \boldsymbol{x}^T \ddot{\boldsymbol{R}}_n \boldsymbol{x} - 2\,\boldsymbol{x}^T \dot{\boldsymbol{r}}_n - \rho_n\,, & q \text{ odd, } n \in \mathcal{I}_{CC} \cup \mathcal{I}_{EI} \end{cases}
$$

$$(3.28)$$

with $q = 1, \ldots, 2N_{CC} + 2N_{EI}$. Then the qth application of the SD algorithm solves

$$\text{find } \boldsymbol{x} \text{ such that } g_q(\boldsymbol{x}) \le 0$$

$$(3.29)$$

$$\text{subject to } g_i(\boldsymbol{x}) \le 0,\ i = 1, \ldots, q-1$$

using the algorithm of Figure 3.2, where gradients of (3.28) are given by

$$
\nabla g_q(\boldsymbol{x}) = \begin{cases} 2\,\ddot{\boldsymbol{R}}_n \boldsymbol{x} + 2\,\dot{\boldsymbol{r}}_n\,, & q \text{ even, } n \in \mathcal{I}_{CC} \cup \mathcal{I}_{EI} \\ -2\,\ddot{\boldsymbol{R}}_n \boldsymbol{x} - 2\,\dot{\boldsymbol{r}}_n\,, & q \text{ odd, } n \in \mathcal{I}_{CC} \cup \mathcal{I}_{EI} \end{cases}
$$

$$(3.30)$$

Step (2): In this case we apply the SD algorithm with objective function (3.15) under all power constraints (3.28), to solve

$$\text{minimize } f(\boldsymbol{x}) = \boldsymbol{x}^T \ddot{\boldsymbol{R}} \boldsymbol{x} + 2\,\boldsymbol{x}^T \dot{\boldsymbol{r}} + \rho$$

$$(3.31)$$

$$\text{subject to } g_i(\boldsymbol{x}) \le 0,\ i = 1, \ldots, Q = 2N_{CC} + 2N_{EI}.$$

Here the gradient is $\nabla f(\boldsymbol{x}) = 2\,\ddot{\boldsymbol{R}} \boldsymbol{x} + 2\,\dot{\boldsymbol{r}}$.

Step (3): In this case we apply the algorithm (3.27) N_{CC} times in cascade, for meeting the power quality constraint on each PCC. The reference algorithm is (3.29), where $q = 2N_{CC} + 2N_{EI} + 1, \ldots, 3N_{CC} + 2N_{EI}$, and where the power quality constraint functions for $q > 2N_{CC} + 2N_{EI}$ are of the form

$$
\begin{aligned}
g_q(\boldsymbol{x}) = \boldsymbol{x}^T[(1 - \epsilon)\ddot{\boldsymbol{D}}_n - \boldsymbol{N}_n]\boldsymbol{x} \\
+ 2\,\boldsymbol{x}^T[(1 - \epsilon)\dot{\boldsymbol{d}}_n - \boldsymbol{n}_n] + [(1 - \epsilon)\delta_n - v_n]
\end{aligned}
$$

$$(3.32)$$

with $n \in \mathcal{I}_{CC}$, and with gradients

$$\nabla g_q(\boldsymbol{x}) = 2[(1 - \epsilon)\ddot{\boldsymbol{D}}_n - \boldsymbol{N}_n]\boldsymbol{x} + 2[(1 - \epsilon)\dot{\boldsymbol{d}}_n - \boldsymbol{n}_n]$$

$$(3.33)$$

Step (4): In this case we apply the SD algorithm of Figure 3.2 with objective function (3.15) as in (3.31), but with all $Q = 3N_{CC} + 2N_{EI}$ constraints (3.28) and (3.32) active.

3.6 Distributed Implementation

In this section we consider a possible distributed implementation of the SD algorithm discussed in the previous section. This can be defined, provided that a simple way of computing gradients is available, since constraint verification is a straightforward task that can be obtained by local measurements and by an acknowledge message. In the distributed implementation, time is divided into slots and at each slot a master assigns to one PEI at a time (in a round-robin fashion) the task to update its injected current.

Step 1: Gradients of interest in step 1 are those referring to local power derivatives. During the minimization/maximization of power P_n to meet its constraints, the gradient of interest is

$$\frac{dP_n}{d\,\Re\left[I_{0,m}^{(k)}\right]} = \begin{cases} 2\Re\left[\left[a_n^{(k)}\right]_{\varphi(m)} V_{0,n}^{(k)}\right], & n \in \mathcal{I}_{\mathrm{CC}} \\ 2\Re\left[\left[b_n^{(k)}\right]_{\varphi(m)} I_{0,n}^{(k)} + \delta_{m,n} V_n^{(k)}\right], & n \in \mathcal{I}_{\mathrm{EI}} \end{cases} \qquad (3.34)$$

with $\delta_{m,n}$ the Kronecker delta. A perfectly equivalent result is obtained for imaginary parts by simply replacing $\Re$ by $\Im$. Note from (3.34) that, apart from local measurements $V_{0,n}^{(k)}$, $I_{0,n}^{(k)}$, ad $V_n^{(k)}$, vectors $a_n^{(k)}$ or $b_n^{(k)}$ must be known. Incidentally, these can be estimated from current/voltage variations (see (3.2))

$$\left[a_n^{(k)}\right]_{\varphi(m)} = \Delta I_n^{(k)} / \Delta I_m^{(k)}, \qquad n \in \mathcal{I}_{\mathrm{CC}}$$
$$\left[b_n^{(k)}\right]_{\varphi(m)} = \Delta V_n^{(k)} / \Delta I_m^{(k)}, \qquad n \in \mathcal{I}_{\mathrm{EI}} \qquad (3.35)$$

e.g., at a proper sounding session at the microgrid start-up. Then the distributed protocol for minimization/maximization of power P_n of PEI or PCC n can be organized as follows:

a. Cycle until the constraint on P_n is met.
b. For $m \in \mathcal{I}_{\mathrm{EI}}$ and $k = 1, \dots, K$:
   i. PEI m sets $\gamma = \gamma_0$
   ii. While $\gamma \geq \gamma_{\min}$
     A. *Inject currents.* Using stored estimates and local measurements in (3.34), PEI m alternatively updates the real and imaginary parts of currents according to[5]

$$I_{0,m}^{(k)} \implies I_{0,m}^{(k)} \mp \gamma\, d, \qquad d = \begin{cases} \dfrac{dP_n}{d\Re\left[I_{0,m}^{(k)}\right]} \\ j\dfrac{dP_n}{d\Im\left[I_{0,m}^{(k)}\right]} \end{cases} \qquad (3.36)$$

---

[5] $j = \sqrt{-1}$.

and broadcasts the new value. The minus sign in (3.36) is for minimization, and the plus sign for maximization of P_n.

B. *Estimate.* Using the broadcasted value and local measurements, PEI and PCC update their estimates of $[a_n^{(k)}]_{\varphi(m)}$ and $[b_n^{(k)}]_{\varphi(m)}$ using (3.35). If the estimates have significantly changed, then the new values are transmitted back to PEI m.

C. *Acknowledge.* PEI and PCC previously meeting their constraints send a NACK when the power constraint is not anymore met at their node.

D. *Confirm.* If no NACK is received and the measured local power has decreased (minimization) or increased (maximization), then the while cycle (b) is concluded.

E. *Cancel.* Cancel the current update, set $\gamma = \gamma \delta_\gamma$, and run the while cycle (b) again.

Note that message exchanges are rather limited.

Step 2: This step is perfectly equivalent to step 1 with the only difference that the gradient is now of the form

$$\frac{dc}{d\,\Re\left[I_{0,m}^{(k)}\right]} = \sum_{n \in \mathcal{I}_{CC} \cup \mathcal{I}_{EI}} c_n \frac{dP_n}{d\Re\left[I_{0,m}^{(k)}\right]}$$

$$= \sum_{n \in \mathcal{I}_{CC}} c_n \Re\left[\left[a_n^{(k)}\right]_{\varphi(m)} V_{0,n}^{(k)}\right] + 2c_m \Re\left[V_m^{(k)}\right] \quad (3.37)$$

$$+ \sum_{n \in \mathcal{I}_{EI}} 2c_n \Re\left[\left[b_n^{(k)}\right]_{\varphi(m)} I_{0,n}^{(k)}\right]$$

Moreover, the external while cycle (1) can be terminated as soon as the cost variation is negligible.

Step 3: This step is perfectly equivalent to step 1 with the only difference that the gradient is now of the form (see (3.33))

$$\frac{dg_q}{d\,\Re\left[I_{0,m}^{(k)}\right]} = 2(1-\epsilon)K_n \Re\left[I_n^{(k)}\left[a_n^{(k)}\right]_{\varphi(m)}\right] - 4\,P_n \Re\left[V_{0,n}^{(k)}\left[a_n^{(k)}\right]_{\varphi(m)}\right]$$

$$(3.38)$$

for $n \in \mathcal{I}_{CC}$ and requires that a local measurement of $I_n^{(k)}$, $V_n^{(k)}$, and $V_{0,n}^{(k)}$ (hence also of K_n and P_n) is available. The confirmation step 4. will include confirmation on already met quality constraints.

Step 4: This step is perfectly equivalent to step 2 with the only difference that PCC must also send a NACK whenever a quality constraint is no more met.

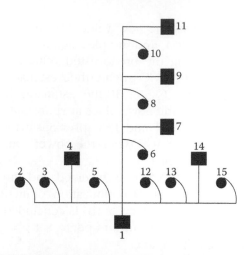

Figure 3.3   Example of microgrid. Circles represent loads. Rectangles represent PEI, except for node 1, which is a PCC.

3.7 Numerical Results

In order to assess the performance of the proposed techniques we consider the microgrid of Figure 3.3 comprising 5 PEI, 1 PCC, and 9 loads.

The power factor constraint has been set to $\lambda \geq 95\%$. Power constraints, cost factors, and electrical parameters of the microgrid are reported in Table 3.1. The set of PEI includes two renewable source generators having zero cost (with node indices 7 and 11), and three PEI with unitary cost. The PCC has a cost that we let vary in order to evaluate the impact of the optimization process for different costs of the energy provided by the utility. The parameters for the SD algorithm are $\gamma_0 = 1$, $\delta_\gamma = 0.7$, and $\gamma_{min} = 3.2 \cdot 10^{-16}$.

The total cost of the microgrid as a function of the cost factor of the PCC is shown in Figure 3.4. The two curves in the figure refer to either the result obtained by the proposed SD algorithm, or to the optimal result that can be obtained with a high computational burden using optimization tools (e.g., GAMS [8]). The quality factor behavior is reported in Figure 3.6, and the powers provided by PEI and the PCC are then given in Figure 3.5.

From Figure 3.4 we see that the SD approach provides a close to optimal solution for the whole range of c_1. Indeed, the total cost of SD exhibits some fluctuations due to the fact that the found solution is in general only a local minimum cost solution, that may differ from the global optimum solution. In Figure 3.5 we see how PEI corresponding to renewable sources are always exploited at their maximum, as expected in an efficient management

Table 3.1   Parameters of the Example Microgrid

	PCC 1	PEI 4	PEI 7	PEI 9	PEI 11	PEI 14
c_n [unit cost/W]	interval [0.5,2]	1	0	1	0	1
U_n [kW]	400	30	30	30	30	30
L_n [kW]	0	0	0	0	0	0
$V_{0,n}^{(k)}$ [V]	[220,10,2,1]	—	—	—	—	—

Load impedances $Z_n^{(k)}$ [Ω] $= 3.385 + j\,2.54$.
Line impedances $Z_{\text{line }n,m}^{(k)}$[$\Omega$] $= 0.06 + j\,0.00194\,k$.

of the microgrid. We also see that, as long as the cost factor of the PCC is lower than the (nonzero) costs of PEI, the PCC provides a significant fraction of power, which decreases as c_1 increases. This is also reflected in Figure 3.4, where we observe that for $c_1 < 1$ the optimal total cost is almost linear with c_1. For higher values of c_1, PEI operate close to their saturation (in terms of provided power) in order to reduce the PCC usage. Finally, note in Figure 3.6 how the quality constraint is always strictly met.

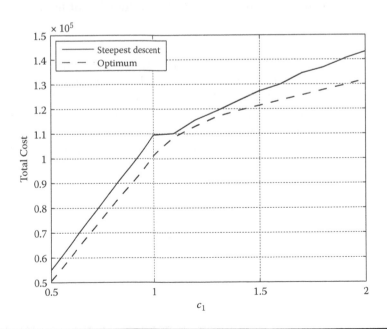

Figure 3.4   Total cost of the microgrid as a function of the cost factor of the PCC.

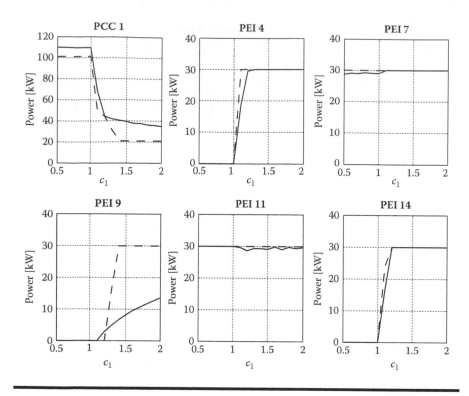

Figure 3.5 Power provided by PEI and PCC as a function of the cost factor of the PCC.

Some insights on the SD algorithm are provided in Figure 3.7 as a function of the iteration number and for $c_1 = 1.1$. In the figure we show the behavior of the total cost, the PCC power, and the quality factor.

Note how the steps of the SD algorithm of Figure 3.1 are clearly identified by vertical dashed lines in Figure 3.7. Since the algorithm is started at $x = 0$, which is a feasible point for power constraints, step 1 is not activated.

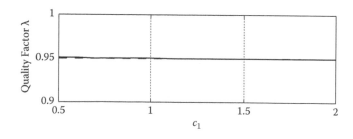

Figure 3.6 Quality factor λ as a function of the cost factor of the PCC.

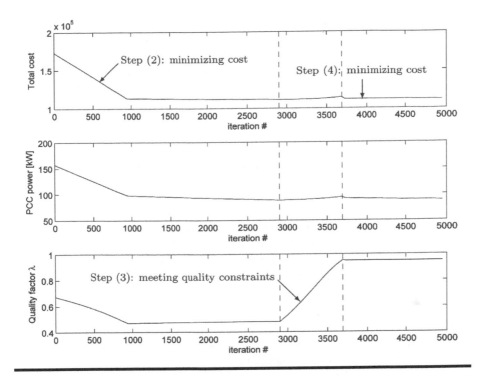

Figure 3.7 **Insights on the SD algorithm for $c_1 = 1.1$. Total cost, PCC power, and quality factor as functions of the SD iteration number.**

In step 2 the total cost is minimized, which implies replacing PCC power with less expensive PEI power. Note how the cost minimization procedure initially advances fast, then slowly optimizes its value. In step 3 the quality constraint is raised to a satisfactory value at the expense of a small cost loss. Finally, cost is minimized in step 4. Note that step 2 reaches convergence in about 1000 iterations, while step 3 requires about 800 iterations and step 4 incrementally decreases the cost.

3.8 Conclusions

In this chapter we have discussed the problem of optimizing the currents injected by DES of a smart microgrid in order to minimize the total cost. Both the optimal and a suboptimal solution based on the SD approach have been considered. A distributed solution suitable for implementation on a power line communication system has also been presented.

Acknowledgments

This work has been supported by the University of Padova within the research project "Design and Implementation of a Novel Control and Communications Architecture for Cooperative Operation of Distributed Harmonic and Reactive Compensators," Prot. CDPA085483, year 2008.

Appendix A: Making the Parameters of (3.2) Explicit

In this appendix we make explicit the relations between $a_n^{(k)}$, $b_n^{(k)}$, $\alpha_n^{(k)}$, and $\beta_n^{(k)}$ in (3.2) and meaningful network parameters.

We preliminarily need to introduce the *network adjacency matrix* A and the *network admittance matrix* $Y^{(k)}$. Let the network have C connections (n, m) between node n and node m, with the request $m > n$ not to count connections twice. Let connections be ordered, with $\mu(n, m)$ the connection number. The adjacency matrix A of size $C \times N$ is defined as a matrix with entries $A_{\mu(n,m),n} = 1$, $A_{\mu(n,m),m} = -1$, and all other entries set to 0. By further letting line impedances be organized in the diagonal matrix $Y_{\text{line}}^{(k)} = \text{diag}(1/Z_{\text{line}\,n,m}^{(k)})$ of size $C \times C$, ordered according to $\mu(n, m)$, the $N \times N$ network admittance matrix is

$$Y^{(k)} = A^T Y_{\text{line}}^{(k)} A \qquad (3.39)$$

stating a relation between node currents and voltages,

$$I^{(k)} = Y^{(k)} V^{(k)} \qquad (3.40)$$

where $I^{(k)} = [I_n^{(k)}]_{n=1,\ldots,N}$, and $V^{(k)} = [V_n^{(k)}]_{n=1,\ldots,N}$. In the following we also let currents and voltages belonging to PCC, PEI, or loads be organized in column vectors

$$
\begin{aligned}
I_{\text{CC}}^{(k)} &= \left[I_n^{(k)}\right]_{n \in \mathcal{I}_{\text{CC}}}, & V_{0,\text{CC}}^{(k)} &= \left[V_{0,n}^{(k)}\right]_{n \in \mathcal{I}_{\text{CC}}}, \\
I_{0,\text{EI}}^{(k)} &= \left[I_{0,n}^{(k)}\right]_{n \in \mathcal{I}_{\text{EI}}}, & V_{\text{EI}}^{(k)} &= \left[V_n^{(k)}\right]_{n \in \mathcal{I}_{\text{EI}}}, \\
& & V_{\text{L}}^{(k)} &= \left[V_n^{(k)}\right]_{n \in \mathcal{I}_{\text{L}}}
\end{aligned}
\qquad (3.41)
$$

node impedances be organized in the diagonal matrices

$$Y_{\text{L}} = \text{diag}\left(\left[1/Z_n^{(k)}\right]_{n \in \mathcal{I}_{\text{L}}}\right) \qquad (3.42)$$

and the network admittance matrix be split in its (nine) components

$$Y_{\text{A,B}}^{(k)} = \left[Y_{n,m}^{(k)}\right]_{n \in I_{\text{A}}, m \in I_{\text{B}}} \qquad (3.43)$$

where A and B may be CC or EI or L. Now, we use relation (3.1) in (3.40), to have

$$
\begin{bmatrix} I_{CC}^{(k)} \\ I_{0,EI}^{(k)} \\ -Y_L^{(k)} V_L^{(k)} \end{bmatrix} = Y^{(k)} \begin{bmatrix} V_{0,CC}^{(k)} \\ V_{EI}^{(k)} \\ V_L^{(k)} \end{bmatrix}
\tag{3.44}
$$

or, equivalently,

$$
M_1 \begin{bmatrix} I_{CC}^{(k)} \\ V_{EI}^{(k)} \\ V_L^{(k)} \end{bmatrix} = M_2 \begin{bmatrix} V_{0,CC}^{(k)} \\ I_{0,EI}^{(k)} \end{bmatrix}
\tag{3.45}
$$

where

$$
M_1 = \begin{bmatrix} -E_{N_{CC}} & Y_{CC,EI}^{(k)} & Y_{CC,L}^{(k)} \\ & Y_{EI,EI}^{(k)} & Y_{EI,L}^{(k)} \\ & Y_{L,EI}^{(k)} & Y_L^{(k)} + Y_{L,L}^{(k)} \end{bmatrix} , \quad M_2 = \begin{bmatrix} -Y_{CC,CC}^{(k)} \\ -Y_{EI,CC}^{(k)} & E_{N_{EI}} \\ -Y_{L,CC}^{(k)} \end{bmatrix}
\tag{3.46}
$$

with E_N the $N \times N$ identity matrix. Therefore, by defining

$$
M = M_1^{-1} M_2 = \begin{bmatrix} M_{CC,CC} & M_{CC,EI} \\ M_{EI,CC} & M_{EI,EI} \\ M_{L,CC} & M_{L,EI} \end{bmatrix}
\tag{3.47}
$$

it is

$$
\begin{aligned}
I_{CC}^{(k)} &= M_{CC,EI}^{(k)} I_{0,EI}^{(k)} + \left(M_{CC,CC}^{(k)} V_{0,CC}^{(k)} \right) , \\
V_{EI}^{(k)} &= M_{EI,EI}^{(k)} I_{0,EI}^{(k)} + \left(M_{EI,CC}^{(k)} V_{0,CC}^{(k)} \right)
\end{aligned}
\tag{3.48}
$$

By comparison with (3.2), it turns out that $(a_n^{(k)})^*$, $n \in \mathcal{I}_{CC}$ are the rows of $M_{CC,EI}^{(k)}$ and $\alpha_n^{(k)}$, $n \in \mathcal{I}_{CC}$ are the elements of the vector in brackets in the first equation of (3.48). Similar observations hold for $(b_n^{(k)})^*$ and $\beta_n^{(k)}$.

References

1. G. Mauri, D. Moneta, C. Bettoni, and G. Colombo. Smart multimetering and dual fuel tariffs for integrating active customers in smartgrids. In *IET Conference Publications*, no. 550(2009).
2. R. Bacher and H. Glavitsch. Loss reduction by network switching. *IEEE Transactions on Power Systems* **3** (2), 447–454 (1988).

3. W. Shao and V. Vittal. Corrective switching algorithm for relieving overloads and voltage violations. *IEEE Transactions on Power Systems* 20 (4), 1877–1885 (2005).

4. R. P. O'Neill, R. Baldick, U. Helman, M. H. Rothkopf, and W. Stewart. Dispatchable transmission in rto markets. *IEEE Transactions on Power Systems* 20 (1), 171–179 (2005).

5. R. Bacher and H. Glavitsch. Network topology optimization with security constraints. *IEEE Transactions on Power Systems* 1 (4), 103–111 (1986).

6. B. G. Gorenstin, L. A. Terry, M. V. F. Pereira, and L. M. V. G. Pinto. Integrated network topology optimization and generation rescheduling for power system security applications. In *Proceedings IASTED International Symposium: High Technology in the Power Industry*, Bozeman, MT, pp. 110–114 (August 1986).
IASTED International Conference On High Technology In The Power Industry

7. B. Wille-Haussmann, T. Erge, and C. Wittwer. Decentralised optimisation of cogeneration in virtual power plants. *Solar Energy* 84, 604–611 (2010).

8. General Algebraic Modeling System (GAMS). http://www.gams.com.

Chapter 4

An Application of Multiperspective Service Management in Virtual Power Plants

Matthias Postina, Sebastian Rohjans,
Michael Specht, Ulrike Steffens, Joern Trefke,
and Mathias Uslar

Contents

Within the energy domain, the manifold term *service* gains more and more momentum. The term is used in different contexts and on various abstraction layers, so that many perspectives on services exist. This chapter introduces multiperspective service management (MPSM) and its application in virtual power plants (VPPs). A VPP combines numerous decentralized generating units and consumers by a smart ICT-infrastructure, and thus creates a need for service-oriented architecture management. Service management has to address several views from different stakeholders from certain viewpoints. MPSM allows both the annotation of metadata to single services to support context-free analyses and modeling the entire enterprise for holistic and context-aware service management. This approach is based on Enterprise Architecture Management knowledge, and it shows how to manage complex systems like VPPs by considering different perspectives on services and domain-specific requirements. To demonstrate the benefits of the approach, exemplary analyses are provided as use cases.

4.1 Introduction

In recent years, the perception of service-oriented architecture (SOA) has undergone a remarkable change. Today, SOA is no longer considered a technical solution combining web services over standardized interfaces and protocols. Instead, SOA is conceived as a conceptual approach in order to align enterprise IT systems with the business strategies and processes they are supposed to support. The notion of service as the lowest common denominator of business and IT has become successful for consolidating IT application landscapes according to business needs. Combining services in ever new service choreographies allows a flexible, easily adaptable, and thus sustainable IT support, and SOA is state of the art for implementing such environments.

As SOA grows in enterprises, management of such service-oriented enterprise architectures is of utmost importance. Today, in a world of global networked virtual enterprises, service management just relying on Universal

Description Discovery and Integration (UDDI) or similar simple techniques seems like playing pool with a rope. Smarter approaches for SOA management have to be considered, especially when large-scale systems are expected, like smart grids in the energy domain. The National Institute of Standards and Technology (NIST) predicts significant changes to the energy sector caused by challenges arising from the electric smart grid endeavor [1]. Utilities as important stakeholders of the smart grid are confronted with drastic changes in business, IT, and technology, so change management on the enterprise level or even beyond will be essential for succeeding in enterprise adoption and interaction with a changing environment.

Modern SOA management has to provide holistic analysis in context of the entire enterprise or even beyond its boundaries. The discipline of Enterprise Architecture Management (EAM) offers profound knowledge in terms of techniques, methods, and frameworks to model and analyze enterprise architectures (EAs) and modern EA frameworks like The Open Group Architecture Framework [2] (TOGAF) or the Department of Defense Architecture Framework [3] (DoDAF) have begun to address service-oriented EA. However, a holistic consideration of entire enterprise service landscapes as part of EAM is still a daunting task for research and praxis.

In Postina et al. [4] we introduced high-level views on SOA in context of smart grid and identified early conceptual mappings of EA-related and energy domain-specific views. This chapter has more practical focus and introduces an application of multiperspective service management (MPSM) in virtual power plants (VPPs). VPP is a concept for the future energy grid combining numerous decentralized generating units and consumers by a smart information and communication technologies (ICT) infrastructure. Within such an infrastructure, many virtual and physical components have to exchange data among each other. So, for example, devices have to be monitored and controlled, but also business processes like accounting have to be carried out. This results in a strong need for various different services that have to be managed as part of SOA.

In the following, we show how MPSM helps to manage a complex system like virtual power plants by considering different perspectives and domain-specific requirements. Two use cases will be introduced applying our approach and enabling exemplary architecture analyses.

4.2 Virtual Power Plant

In the last few years, the domain of power generation has undergone major changes, especially in terms of distributed generation. The generation share of decentralized energy resources (DERs) has increased enormously [5]. This increase is largely owing to the efficiency enhancement in existing DERs and the development of new technologies like fuel cells, tidal generators, and small hydro generators, being complemented in the near future by plug-in electric vehicles (PEVs), which can increase the storage capacity.

Main drivers of this transformation to a distributed power grid are ecological and political reasons as, for example, the European Union targets to replace 20% of the centralized power supply with DERs by 2020 [6].

To handle and integrate the emerging amount of DERs and the newly arising prosumer[1] role, an adaptive power system is needed.

The concept of virtual power plants (VPPs) is a strategy to integrate the new stakeholders of the future energy grid [7]. External parties just observe the VPP as a single entity similar to classical power plants, hiding the internal structure. This internal structure is far more complicated and demands an incisive information and communication technologies (ICT) infrastructure to establish a centralized control over the various DERs, controllable loads like cold storage warehouse and energy storage devices like pump storage units or PEVs. A VPP thereby has the potential to combine the advantages of various renewable energy sources. Wind turbines and solar modules generate electricity according to the weather. Biogas and combined heat and power (CHP) units as well as available energy storage devices are used to make up the difference between generation and demand. However, to accomplish this, various steps are needed.

First, it is necessary to distinguish between anticipatory control (day-ahead planning) and fine-tuning similar to final system balancing at the time of delivery (real-time balancing or reactive action planning).

The day-ahead planning is based on forecasts for the demand and production. Demand profiles are created on the basis of "load profiles" that are based on metered history consumption, whereas the production forecasts are generated with the aid of weather forecasts. Because wind and solar energy cannot precisely meet a given electricity demand, oversupplies and shortages need to be compensated by storages as well as biogas and CHP units. The forecasts help to generate day-ahead schedules for these units (see Figure 4.1).

Even with high-quality weather forecasts, the actual production of wind and solar power, as also the real consumption, can differ from predictions. To deal with this, real-time balancing based on the actually measured values has to be done on the operating schedules (see Figure 4.2).

Another possibility is to take active control of the demand, including customers, to influence the amount and the timing of electricity usage. This new paradigm is called demand-side management (DSM) and usually involves advanced information technology [9].

---

[1] Prosumer is a portmanteau word originating from the two words *producer* and *consumer*. In the energy domain it means that a prosumer can be a utility's client in terms of consumption as well as a supplier in terms of production.

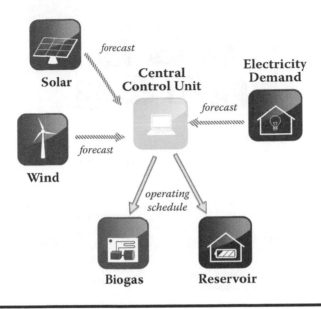

Figure 4.1   Day-ahead planning based on [8].

The maintenance of the balance between demand and production is not the only option for a VPP. From the economical viewpoint it makes sense to sell the surplus energy or to even increase the energy production in high-priced times. The bundling of enough small-sized DER and greater

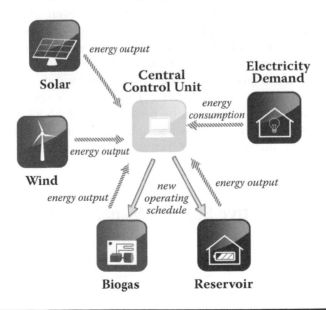

Figure 4.2   Real-time balancing based on [8].

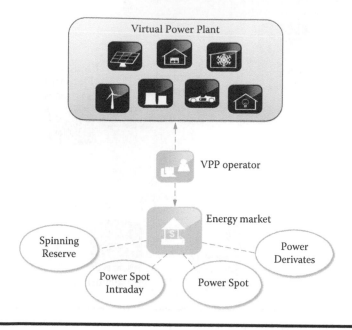

Figure 4.3 VPP market connection.

production units like wind and solar parks makes it possible to participate in energy markets with entry barriers such as the European Energy Exchange (EEX). The VPP operator has the option to offer several energy products like spinning reserve, intraday power products, or even spot power products (see Figure 4.3). To optimize the ecological and economical result, it is necessary to consider different boundary conditions such as the technical aspects of the DER and storages, the electrical and thermal demand, the inaccurate prediction of wind and solar energy production, and furthermore the prognosis of energy market prices. This leads to a very complex optimization problem, which can only be resolved with high information flows between different stakeholders inside and outside the VPP, and therefore an advanced ICT infrastructure is needed to ensure the communication paths.

4.3 Multiperspective Service Management

The term service-oriented architecture (SOA) has been out there for about a decade now—joined by confusion about its definition and characteristics ever since. Causes for confusion were diverse, changed over time and will probably exist as long as such paradigms pursue Gartner's hype cycles. However, we are not going to cut this Gordian knot here. We rather focus on different stakeholder perspectives involved in SOA as one cause for misconception and discuss our approach of multiperspective service management (MPSM) to harmonize viewpoints.

A multiperspective SOA management approach is imperative, since many enterprises are adopting or have already adopted the paradigm of SOA to align business and IT. Following this paradigm, services seem to become the central building blocks of entire enterprise architectures with multiple stakeholders involved. This trend changes the characteristics of enterprise architectures, since the subject matter is shifting from applications toward services and leads to a number of new challenges, like:

- Applications were considered as black boxes with hidden internal functionality offering a limited number of interfaces to the environment. Internal integration was generally assured by data integration on a common database. In SOA-like environments also former internal functionality is made available externally for multiple purposes. This leads to a new level of granularity, the need for a formalization of interface descriptions, and explicit data transfer between services. Also, formally hidden aspects like security, transaction and error handling, and assured SLAs need to be externalized.
- Applications were usually reflecting organizational structures. A billing application was managed and used by the billing department, for instance. This led to a natural alignment of systems and responsibility. In SOA, environment services can be used in various contexts to assemble temporary composites. Responsibilities need to be modeled explicitly in such environments.
- The possibility to exchange services easily fosters the development and usage of alternative services. Functional and nonfunctional requirements need to be clearly stated by all involved stakeholders in order to find an appropriate match.
- Applications were also logical entities with natural borderlines for architectural rules and standards. Protocols, data models, technology specification, as well as version and release control were bound to an application or at least modules of an application. Service management needs to support these issues for much more and smaller entities.

This list reflects only some of the challenges arising from the architectural paradigm shift but should be sufficient to motivate the need for efficient enterprise service management. Before we consider the MPSM approach in detail, we would like to introduce some terms and definitions.

4.3.1 Terms and Definitions

MPSM is our service management method for large service-oriented enterprise architectures. It is inspired by techniques used in the discipline of

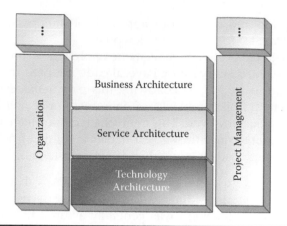

Figure 4.4   Enterprise architecture overview [4].

enterprise architecture (EA) and adopts its management perspectives used in classical application landscape management. This section introduces terms and definitions for the architectural description of service-oriented enterprise architectures, particularly with regard to the development of perspectives for service management.

4.3.1.1 Enterprise Architecture

According to the Enterprise Architecture Research Forum,[2] enterprise architecture can be characterized as "the continuous practice of describing the essential elements of a sociotechnical organization, their relationships to each other and to the environment, in order to understand complexity and manage change." A sociotechnical organization could be a governmental agency, an enterprise, a part of an enterprise, an enterprise with customers and suppliers, or an entire industry sector, for example. EA is able to describe the current state of a sociotechnical organization and also to plan envisioned states. EA as a discipline has developed tools, methodologies, and frameworks to manage such changes and offers models and a common vocabulary to formalize the essential elements and their relationships to each other and to the environment.

EA considers various architectures of a sociotechnical organization and provides adequate views to different stakeholders. The center of Figure 4.4 shows typical architectures of a sociotechnical organization as well as other EA aspects, like project management, organizational aspects, security, etc. Each entity of Figure 4.4 stands for an abstraction of a sphere of interest of different stakeholders.

---

[2] http://earf.meraka.org.za/earfhome.

A business analyst, for example, is interested in the optimization of business processes. He has specific concerns like decomposition of work flows, runtime of tasks, or costs associated to business processes. A business analyst is used for specific notations like Event-driven Process Chain (EPCs) or Business Process Model and Notation (BPMN) diagrams and performs a specific set of analyses on a stakeholder-specific information model. Obviously, the business analyst is interested in the business architecture of an organization. A top manager may also be interested in the business architecture, but has different concerns and is used to certain reports and visualizations. To provide a coherent picture of the business architecture, helping stakeholders to understand complexity and manage change, EA has to model the essential business elements, their relationships to each other and to the environment—an abstraction of the business sphere of interest.

Besides business architecture, EA provides abstractions to other spheres of interest, like technology architecture and service architecture, and considers organizational structures and projects of a sociotechnical organization and interconnects these partial models to a coherent content model. Such model weaving fosters the real value of EA, since it leads to a consistent and comprehensive representation of a sociotechnical organization and allows complex analysis over all spheres of interest. The entire model is typically stored inside a centralized EA repository to enable access to all relevant stakeholders.

The model stored inside a repository is organization specific and depends on stakeholder concerns. However, patterns for typical concerns exist (Buckl et al. [10], Gringel and Postina [11]), as well as generic content models appropriate for a number of enterprises from different sectors (see TOGAF [2] or DoDAF, [3] for example). Such content model templates comprehend EA best practices and ease the development of a tailored and organization-specific model.

Another important aspect of using content models and standardized EA frameworks lies in the common language of enterprise architects. A common understanding of things, a clear taxonomy describing essential entities, and a model showing the relationships of entities to each other enable a semantic description of the EA discipline, and moreover allow consideration of EA in context of other disciplines.

4.3.1.2 Business, Application, and Infrastructure Services

SOA relies on services as central building blocks for entire enterprise architectures. Unfortunately *service* is an ambiguous term and demands closer consideration. For MPSM, we follow the definition of business, application, and infrastructure service provided by ArchiMate [12].

SOA was introduced as paradigm to build a bridge between business and IT departments. Closing this mental gap is a challenging task, but the notion of service as a common term seems to help business analysts to model business activities inside of processes as requests for corresponding IT service support. "Record customer data" is an example of such business activity, which has to be supported by IT by providing data forms, for instance.

Standing on this bridge and looking in both directions, business and IT, we can begin our consideration with the separation of business services from automated services. The business activity "Record customer data" could be part of a selling process, which is a service of the enterprise to a customer. Regarding the IT direction, it specifies the demand for IT support in terms of an interaction service offering a data form to a call center agent, for instance.

Following this interaction service downward, we will discover multiple other IT services supporting this interaction service. The data entries of the call center agent might be checked automatically to ensure a correct zip code before the data are stored by another service into the database of some Customer Relationship Management (CRM) application. A credit check might also be performed simultaneously as part of this business activity or is triggered later on.

The automated services are deployed on application components like central process orchestration engines, for example, or on CRM, billing, or other applications. Especially when business processes have to change often, large enterprises establish central process engines where services can be orchestrated flexibly to ensure aligned workflow support. We call these kinds of automated IT services application services. Application services rely on infrastructure services like processing, storage, and communication services, which are deployed on infrastructure devices like servers.

Figure 4.5 shows three layers of services: the business layer hosting business services as part of business processes or activities, the application layer, and the infrastructure layer. Business services as indicated by sketched business process models are provided and consumed by certain business roles. Automated IT services are indicated by boxes with white chevrons—the area between the gears symbolizes application services, and the area between the cpus infrastructure services. These services form hierarchically layered building blocks of the enterprise architecture and are considered the service landscape of an enterprise.

The general meaning of the black arrows is a "uses reference" between services. So business services can use application services, which means that business activities are supported by application services. These application services use other application or infrastructure services, etc.

4.3.1.3 Stakeholder, Concerns, View, and Viewpoints

As in every complex IT environment, many people are involved performing various roles when service-oriented environments are planned,

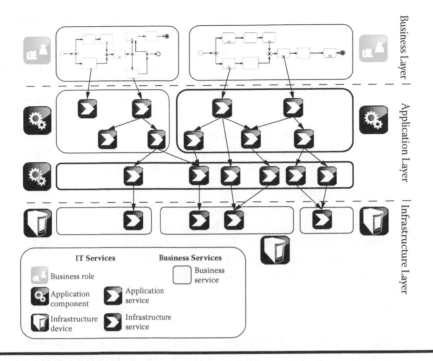

Figure 4.5   Services in enterprise architectures.

implemented, and run. As SOA stakeholders have different concerns, ad-equate views on the service landscape are necessary to satisfy these con-cerns. Services of the service landscape, for example, can be considered as building blocks for higher-order functionality, which can also be referred to as business capabilities of an organization. The architecture management is particularly interested in these groups of services and regards them as com-posite or logical applications on a more abstract level, neglecting technical details of implementation.

Another view might be related to the role of a developer, for whom tech-nical details are of major importance. For developers, issues like interface descriptions, response times, pre- or postconditions, and error handling matter, as well as mechanisms for service discovery or life cycle information.

We regard a service-oriented enterprise architecture as a complex sys-tem of systems and follow the terms of *IEEE Recommended Practice for Architectural Description of Software-Intensive Systems* (IEEE 1471 [13]) to describe stakeholder, concerns, view, and viewpoints. However, to obtain a detailed overview, especially to relate to concepts omitted in these con-siderations, we refer the reader to the IEEE standard.

The service-oriented enterprise (system) is relevant for various stake-holders and has an architecture described by an architectural description. Furthermore, an architectural description selects one or more viewpoints for use. What viewpoints are selected is typically based upon considering the stakeholders to whom the architectural description is addressed and their concerns. The architectural description is organized by views conforming to a specific viewpoint whose definition either stems from the architectural description itself or may have been given elsewhere. In the latter case, the viewpoint is referred to as a library viewpoint. A view may be composed by one or more architectural models, and each model is developed using the methods established by its associated architectural viewpoint. An architectural model may participate in more than one view. According to the IEEE standard, each viewpoint shall be specified by:

■ A viewpoint name
■ The stakeholders to be addressed by the viewpoint
■ The concerns to be addressed by the viewpoint
■ The languages, modeling techniques, or analytical methods to be used in constructing a view based upon the viewpoint
■ The source, for a library viewpoint

We adhere to these recommendations and specify all our SOA viewpoints accordingly. Table 4.1 shows the exemplary viewpoint *Process Support* following the aforementioned specification. As also proposed in Postina et al. [14], the viewpoint has a name, the stakeholder addressed is mentioned and the underlying concern is described by the fields "Question," "Description," and "Intent."

Enterprise architecture makes extensive use of visualizations to support analysis by visual views. Whenever possible, we describe the modeling technique to create the corresponding views by referring to one or more software maps (as defined in Wittenburg [15]). We also identify "Required Data" entities necessary to design the metamodel. Finally, if applicable, the source for a library viewpoint is indicated by the field "Viewpoint Source." In addition to the IEEE recommendations, we include the field "Data Source" in our viewpoint specification. This is important in order to estimate the effort for data collection—if one can extract needed data automatically from a process engine, for example, the effort to make viewpoints available to stakeholders can be radically reduced.

4.3.2 Perspectives

The service landscape needs to be structured to allow a stakeholder-specific classification of services. Such landscape partitioning appears in different ways, and literature provides various categories for service classification (see Erl [16], Krafzig et al. [17], Josuttis, [18] or Engels et al. [19] for various classification approaches).

Annotating services with metadata is one common way to specify services. Information about the service provider, interface descriptions,

Table 4.1   Viewpoint Example—Process Support

Name	Process Support
Question	Which business process is supported by which application services?
Description	Business processes usually consist of multiple activities and are automated by application services. All services and their contributions to business processes or to business activities, respectively, should be determined for this concern.
Intent	Due to the overview given, the current alignment of business and IT can be examined in more detail. It can, for instance, be checked to which degree critical business processes are supported by services or which processes/activities offer more potential to be automated by services. It can also be detected, whether a service contributes to a business process or not (e.g., architecture violation). The information yielded by this concern can be used as a planning basis for service landscape evolution.
Stakeholder	Business architect
Visualization	Dependency graph, process map (matrix map)
Required Data	*Business process, business activity, business service*
Viewpoint Source	Pattern catalog [10] (V-18), project specific
Data Source	Process engine

descriptions of SLAs, etc., is specified and stored as metadata for a corresponding service. Techniques like Universal Description, Discovery, and Integration (UDDI) for service registries, the Web Service Description Language (WSDL), or semantic annotation techniques like Web Service Semantics (WSDL-S) and the Web Ontology Language for Web Services (OWL-S) were invented for service description.

Such service annotation techniques focus on the single service in order to support service discovery and integration. The description is bound to the service and can be regarded as a context-free perspective on services. It is context-free in so far as context information like typical use cases of a service is not regarded. So information about who uses the service, what processes rely on a certain service, or what actual infrastructure is used by dependent services can not be derived by context-free service consideration.

Considering deployment and orchestration descriptions to externalize calling hierarchies of services adds such context information. Moreover,

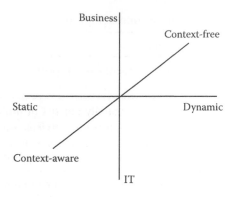

Figure 4.6   Mental framework for viewpoint classification.

combining service dependency knowledge with enterprise architecture knowledge allows context awareness of services and enables stakeholders to perform deeper architectural analysis, as we will see in Section 4.3.4. For now, both perspectives, context-free and context-aware service consideration, become part of our mental framework, as shown in Figure 4.6.

The y-axis of this framework divides business and IT perspectives on services. As already discussed in Section 4.3.1, information on services is more or less important for different stakeholders. Developers have other viewpoints on architecture than business analysts, to give an example.

Within our mental framework, we distinguish further between dynamic and static information about services (Figure 4.6). Static information is gathered at design time; dynamic information is the result of runtime analyses like monitoring information such as the frequency of service usage, error rates, and investigation of services never used.

4.3.3 The Metamodel for MPSM

The goal of MPSM is to provide-adequate information about the service landscape in terms of stakeholder-specific viewpoints to support stakeholders decision making. Due to its complexity, such a task can not be performed manually, so adequate tool support is required.

Section 4.3.2 introduced a mental framework hosting various perspectives to provide a rough coordination system for classification of service management-related viewpoints. However, to set up an information system able to support MPSM, a more formal approach should be introduced in terms of a metamodel able to support the introduced perspectives.

Figure 4.7 shows a quite generic metamodel similar to Ecore (see Steinberg et al., [20] for example), which indeed inspired us. One major difference is that we use references as first-class elements in order to use

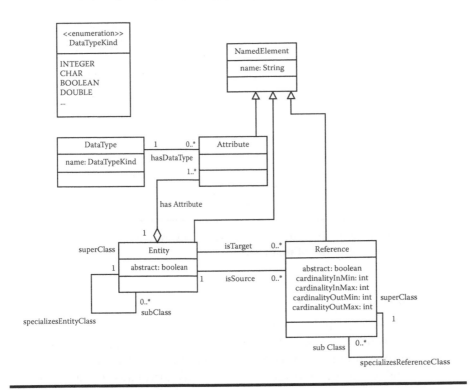

Figure 4.7   Metamodel of MPSM.

the full power of inheritance also on references. The metamodel allows us to model graph-like structures and refinement through inheritance hierarchies on entities and references. Figure 4.8 shows such refinement. On the left-hand side we see the refinement of entities. We separate between service and nonservice entities, where services reflect the structural part of the service-oriented enterprise architecture and nonservice entities are used to model the context. Refinement of references is illustrated on the right-hand side. Source and target describe restrictions for entities as valid association ends. For MPSM analysis, we just rely on the entities *EA-Entity, Service,* and *NonServiceEntity,* as well as the references *EA-Relation* and *Uses* as mandatory generic classes (indicated by gray backgrounds in Figure 4.8). All other classes of the content model of an enterprise have to inherit from these super classes.

We would like to use the modeled entities and relations from Figure 4.8 by giving an instantiation example of the model in Figure 4.9. Here entities are depicted as icons, references are symbolized by edges between icons, straight edges stand for a "uses" reference, and dashed edges for the "isResponsibleFor" reference between actors and EA entities.

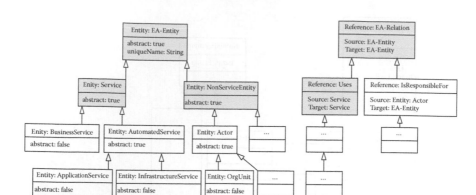

Figure 4.8 Refinement of entities and references by inheritance (gray background: mandatory elements in MPSM models).

4.3.4 MPSM Analysis

In order to provide valuable support to certain stakeholders, the enterprise architecture model needs to be tailored according to stakeholder-specific needs. For the structural part of SOA, this can be done by refinement of the service entities and the introduction of service attributes. Good modeling practice shows that it is not advisable to model all relevant service aspects as attributes on services. Especially data having a service independent life cycle should be specified as single EA entities. These entities provide context to services by meaningful references to services and other EA entities. Figure 4.8 can be seen as a starting point for such model tailoring.

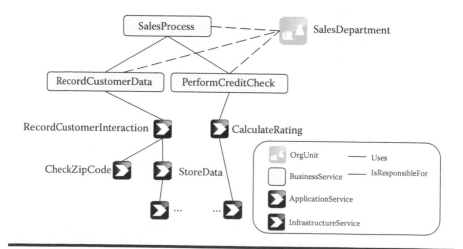

Figure 4.9 Instantiation example.

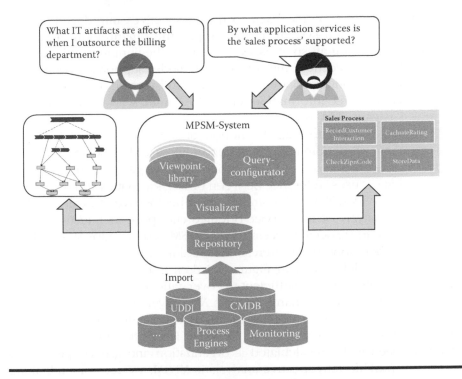

Figure 4.10   MPSM system and typical usage scenario.

Automated data import is a key to success when MPSM is established, since manual data collection is cost- and labor-intensive and also an error-prone task. Normally, various information systems are already implemented in enterprises and can be used as data sources. Service registries and repositories, monitoring systems, Enterprise Service Bus (ESBs), process engines, and Configuration Management Database (CMDBs) serve well for data extraction. We propose the utilization of EA repositories to host the enterprise-specific model and as a place for data imports. Figure 4.10 shows such a repository being part of an MPSM system.

Figure 4.10 also shows the typical use case for an MPSM system. Two stakeholders are using the information system with a certain concern in mind. The system is organized by viewpoints, so stakeholders are able to search for existing viewpoints or can specify new viewpoints based on the enterprise-specific model.

Visualizations have proven to be of value in context of EAM, especially to support fast and intuitive data analysis. The MPSM system provides a standard set of visualizations to support management decision making. Figure 4.10 shows a dependency graph on the left- and a cluster map on the right-hand side. A good MPSM system allows stakeholders to store, reuse, and share viewpoints and configured analysis.

We would like to come back to the perspectives introduced in Section 4.3.2. The MPSM system has to support the organization of various viewpoints reflecting all six perspectives. At the beginning of this section, we already discussed how the information model supports separation of context-free and context-aware information. Attributes on services are used to model context-free information related to a single service, while context awareness is gained by services being part of the entire enterprise model.

Following ArchiMate [12], the notion of service is used in MPSM as the least common denominator between business and IT perspectives. We model business processes and corresponding activities as business services and introduce a "uses" reference between business activities and supporting application services. This link combines both perspectives and allows comprehensive analysis of the service-oriented enterprise architecture. However, a stakeholder-specific access to the MPSM system helps stakeholders to focus on their spheres of interest (see Section 4.3.1.1).

The metamodel shown in Figure 4.7 does not include a concept for temporal modeling. This is sufficient to cover static design time information, but inadequate to cover dynamic aspects. Whenever possible, we aggregate runtime information to a scalar classification and add attributes like business criticality class, frequency of usage class, etc. How often these aggregated attributes need to be recalculated is organization and aspect specific. To be able to provide retrospective analysis in MPSM, we recommend taking snapshots to keep historical information.

4.4 Service Management in VPPs

Virtual power plants (VPPs) can be understood as a federation of multiple individuals or organizations that contribute to a virtual enterprise (VE) in order to achieve a common goal collectively. In this case, virtual power plants aggregate multiple (distributed) power generation units to deliver electric power products that could not be delivered if they act as individual units. Individual power generation units could, for instance, not participate on energy exchanges in order to maximize their benefits. The organizational form of these power plants can be referred to as virtual enterprise (VE). Travicia [21] defines a virtual enterprise, or to be more precise, a virtual organization (VO), as follows:

> VO refers to a new organizational form which manifests itself as a temporary or permanent collection of geographically dispersed individuals, groups, organizational units—either belonging or not belonging to the same organization—or entire organizations that depend on electronic links in order to complete the production process.

As virtual power plants have to coordinate their dispersed production process—as also stated in the preceding definition—they depend on electronic links. To collaborate effectively, there is a strong need and degree for information technology to interconnect the various partners. The collaboration within a VPP will be realized by highly automated services. These information and communication technology (ICT) supported services can be mapped to software services that encapsulate the underlying physical devices' functionality or, in particular, their sensors and actuators. Having a standardized way to access these services allows an effective integration with standard components and a more flexible definition of production workflows. Partners provide services to others (e.g., weather forecasts, production history, sensor data, etc.), which results in a large service network with a variety of services. In order to achieve higher-order goals, these services must be coordinated in a useful way. A prerequisite to successfully compose or orchestrate services is information about their respective functionality, and to integrate with other services, knowledge about the service portfolio is necessary. Gathering this information and deciding on the development or usage of services requires management. To manage the services as a whole, EAM offers appropriate management methods to exploit the organization's full potential.

The management of enterprises or enterprise architecture in particular is the subject of much research and has been evaluated in practice for a long time (see Section 4.3.1.1). Characteristics of VPPs are a high degree of automation and geographic dispersion, which is not explicitly considered in most EAM frameworks, as they primarily address the upper business and application layers. Insofar as VPP management requires these characteristics to be considered, a more elaborated infrastructure layer must be introduced and location information must be made available. Furthermore, time-critical services must receive special consideration, as they might determine grid stability and so can potentially endanger it.

One possible structure to describe the EA infrastructure layer for VPPs in more detail is provided by the IEC 62264 standards family [22]. The standard series is about enterprise control system integration, commonly known as computer integrated manufacturing (CIM). It contains consistent models and terminology for the automation domain in enterprises. In particular, it depicts interfaces between manufacturing control functions, which could be named as infrastructure in terms of EA, and other enterprise functions (business in terms of EA) to resolve language barriers between the different levels of the enterprise and allow integration. At this, the standard introduces a five-level model separating the manufacturing enterprise in a functional hierarchy, which is based on the Purdue hierarchy model [23]. These different levels are business planning and logistics (level 4), manufacturing operations and control (level 3), and batch, continuous, and discrete control (levels 2, 1, and 0).

Based on these levels given, we adopted an automation pyramid from Katzke [24] containing abstract layers decomposing the automation domain in a functional way. The concepts used herein are to be used to classify services in virtual power plants in order to allow service management. The levels depicted as a pyramid are to support the hierarchical understanding from business to technology, and thus show a top-down approach to automation. They assume that commands on top are cascaded to the lower levels with increasing technical nature and automation. Insofar as they are reminders of the common concept of hierarchical command chains in SOA, as well as the functional structure of enterprise architecture, the enterprise is approached from a process-oriented view using applications/services to a more technical level, namely, the physical realizations on hardware, although in a different level of detail.

In order to classify the services and technical aspects in a VPP within a coherent schema, we have considered four layers that classify the automation domain (see Figure 4.11). Each VPP service can be assigned to one of these levels so that rough statements about the functionality about services can be made. Having this context also allows further assumptions to be made for each layer, as services, for example, might only be allowed to initiate communication with services in the same or a lower layer. The classification defined by these layers is defined in the following text (from top to bottom):

Enterprise level: The enterprise level comprises tasks that are not directly concerned with the operational production, but with resource planning. Its activities are concerned with coarse production planning, procurement, and order processing. The tasks on this level are mainly business related and there is no immediate visibility of operational processes.

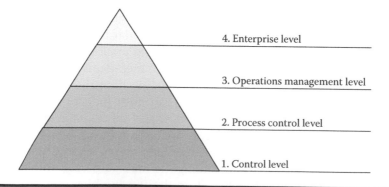

Figure 4.11   Levels of automation---automation pyramid.

Operations management level: Tasks on this level are related to the execution of the manufacturing process. This level is concerned with finer-grained production planning, product data acquisition, measurement of operational performance, and management of materials. Moreover, tasks to manage quality of products and processes are assigned to this layer.

Process control level: This level groups tasks that are related to the production process. Tasks on this level are concerned with supervision, control, and monitoring of the manufacturing process, the management of product plans and their execution. Furthermore, the documentation of indicators for performance and quality as well as measurement are tasks assigned to this layer.

Control level: Within this level, active controls as well as feedback controls carry out the production process. In particular, this functionality is provided by programmable logic controllers (PLCs), which provide the highest abstraction that is considered in our model.

Each of these layers is supposed to have special characteristics, which are important properties for services within this layer. It is assumed that a common set of properties, which are applicable to all services located in this layer, can be identified or defined. These properties then serve as criteria for services retrieval, which is important to analyze and manage services.

With these layers at hand, abstractions for a more detailed description of the technical EA level are given. Services in virtual power plants can then be assigned to one of those layers, which basically build the framework for efficient services retrieval essential for management purposes.

4.4.1 Stakeholders in VPP

In defining a management approach for virtual power plants, it has to be clear which stakeholders are in the focus of this approach. As described in the MSPM approach, stakeholders provide the starting point for the identification of relevant concerns. These can then be incorporated into viewpoints that can be supplemented with appropriate models.

Defining stakeholders is always an organization-specific task, but to simplify the initial identification we identified some common stakeholders that can serve as a starting point. However, there can be several other stakeholders—either internal or external—with interest in VPPs in general, or enterprise architecture or service-oriented concepts in particular. Some stakeholders we identified are characterized in short in the following:

Sponsor: A sponsor initiates programs/projects and ensures their funding. He determines the scope of a program/project and the available financial budget.

Program/project manager: Program/project managers are leading programs or projects and are responsible for the time and budget used within a program or project.

Business engineer: Business engineers must understand business and regulatory requirements and transform them into useful models describing the business functionality (e.g., design and refine business processes). They are also responsible for the validation and maintenance of these models and have to keep track of the corresponding business requirements.

Application engineer: An application engineer has to understand business requirements and to derive appropriate ICT application requirements. Based on these requirements, the choice for concrete applications or their development has to be performed. The application engineer has to describe and validate models based on these requirements. Moreover, he has to continuously ensure the currentness of these requirements and models and communicate respective changes.

Data engineer: Data engineers are responsible for requirements concerning the data needed throughout the application/service landscape. Based on the business and application requirements, corresponding data models have to be defined or chosen by the data engineer and a validation of these has to be performed. Furthermore, the assurance of up-to-date requirements and appropriate data models has to be done by this stakeholder. A data engineer could, for example, identify the common information model (CIM) as a standardized data model and decide on the specific parts to be used.

Technology engineer: A technology engineer has to identify or develop technologies that are necessary to realize the identified business and application functionality. Technology engineers from multiple disciplines may be necessary as, for example, electrical engineers, mechanical engineers, or IT specialists. They also have to continuously validate their chosen technologies and inform other experts about changes.

Security engineer: Security engineers have to develop requirements concerning security, safety, and privacy to ensure a system operation taking these aspects into account. These requirements have to be communicated to all involved stakeholders and have to be continuously checked and validated.

Operations manager: Operations managers are responsible for systems operations and may have interest in various architectural aspects, i.e., descriptions of other experts, especially necessary for troubleshooting.

These stakeholders have concerns regarding the VPP's architecture. On this basis, viewpoints can be identified, which are exemplarily shown in Section 4.4.3. An information model constitutes the basis for organization of the services to be considered.

4.4.2 Information Model for VPP Management

In Section 4.3, the foundations for service management in the context of enterprise architecture were laid. To successfully apply the MPSM method to virtual power plants, the specifics of VPPs covering the management concerns have to be incorporated into an adapted information model. Besides the mentioned layers as structuring elements, we identified some exemplary layer-specific characteristics that are required to allow management based on stakeholder concerns. These VPP specifics are presented in the following text according to the corresponding layer.

4.4.2.1 Enterprise Level

Services on this level are concentrated in a central place and usually not time critical. The subjects of services on the enterprise level are coarse grained and characterized by a low technical detail level and rather a high level of functionality. Thus, we identified the following specific properties: *organization unit* and *automated service execution*.

Organization unit: As the functionality is being defined and used by organization units and services can be used across different organization units, it is inevitable to note the organization unit that is in charge for a service. This information especially becomes necessary, when functionality is to be adapted or new functionality is to be introduced, which may, for instance, be required by other organization units.

Automated service execution: As services on this level are characterized by a high level of functionality, entire processes or composed services may not be fully automated and human intervention may be necessary. Noting if a service is automated or not allows us to identify potential for optimization and to realize timing problems.

4.4.2.2 Operations Management Level

On this level, applications like energy management systems (EMS) or distribution management systems (DMS) were used to fulfill the needs. To support the established understanding of these applications, which are now replaced by services, and to allow for available product integration, this characteristic can be considered. Moreover, these (aggregated) services in this area can run on different, distributed integration platforms, which have to be considered.

SIA application domain: Services on the operations management level can usually be categorized by their application area. It is important to know in which application context a service is used, for instance, if monolithic and complex systems like SCADA will be disentangled on the service level, as will be done in the future to enable SOA. Taking the IEC TR 62357 Seamless Integration Architecture (SIA) as a framework, services can be associated to an application domain according to the SIA. The defined application domains are supervisory control and data acquisition (SCADA), energy management system (EMS), distribution management system (DMS), market operation, engineering and maintenance, and external IT.

Integration platform: Services on the operations management level may be run on different integration platforms as other services, too. To integrate these services successfully, knowledge of the underlying integration platform is required, as these often differ.

4.4.2.3 Process Control Level

The supervision and control of the production process take place on this level. Changes and observations on this level can be immediately concerned with the production, and so the access to services on this level should be well controlled. As services can be composed flexibly, functional blocks can be introduced to group services that are closely related in terms of their functionality. To classify services on this level, the following properties are identified:

Functional domain: Services on the process control level can be assigned to functional domains that describe a group of functions within the VPP. These functional domains can, for example, be control, monitoring, or measurement.

Access level: Access to services on the process control level is often restricted. Access level subsumes rights needed to execute functions offered by the service. Knowing and setting access levels for services allows troubleshooting in case of denied access or to secure the system.

4.4.2.4 Control Level

Services on the control level will be widely distributed and, in our model, they are the finest unit of abstraction. Due to the distributed nature, it is important to consider the location as well as temporal effects.

Location: On the control level, services are usually deployed on geographically distributed devices. Due to this distribution, explicit
knowledge of the location is necessary to assess system risks, for
maintenance and geographic dependency analysis.

Real-time capability: It is necessary to know whether services are real-
time capable. This may especially become necessary if services are
concerned with safety-critical functions and might lead to instability.

An information model, comprising the automation layers and these
properties per layer, is the basis for data and further analyses. A specification of this information model is shown in Figure 4.12. The information
model is based on the MPSM metamodel introduced in Section 4.3.3 and
includes all mandatory MPSM superclasses as basis for inheritance. *EA-
Entity* is root for all entities differentiating further between service and

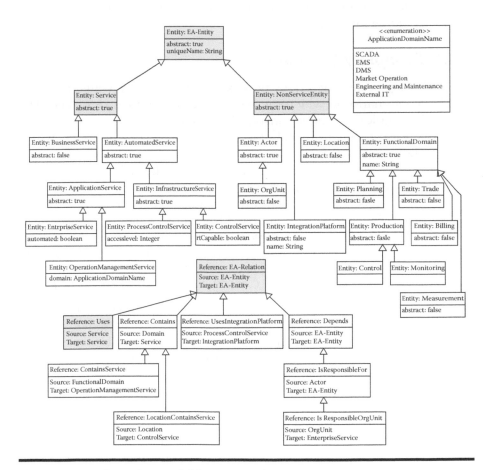

Figure 4.12 Information model for VPP service management.

non-service-related entities. Services then are categorized into abstract services according to the EA layers *Application* and *Infrastructure*. Concrete assignments of services can be made service categories, which are introduced in accordance with the automation pyramid layers. In particular, these services are *Enterprise Service, Operation Management Service, Process Control Service*, and *Control Service*. Furthermore, these services have specific attributes according to the identified properties per layer (e.g., real-time capability for control level services or automated for enterprise level services). Non-service-related entities are *Organization Unit, Location, Integration Platform*, and *Functional Domain*. The abstract *Functional Domain* is further divided into the five categories: *Planning, Production, Trade, Measurement*, and *Billing*, with some exemplary subdomains.

References then primarily serve to associate service entities with non-service-related entities as, e.g., assign a location to the selected service category *Control Service*.

In the following text, we will show two exemplary analyses based on instances of this information model.

4.4.3 Exemplary Use Cases

The following section describes two exemplary use cases within a VPP to build up a scenario on which the concepts can be further described. The selected examples are slightly inspired by Winkels [25].

A main use case can usually be split up into several activities, which have to be completed for the success scenario of a use case. This composition of use cases from activities can be well reflected by services. The functionality and workflow needed for the use cases are realized by services on different levels. Services themselves can use other services to provide their functionality and so be composed from more primitive services. This service call structure then resembles a hierarchical graph, and thus is in accordance with typical SOA-like architectures that are covered by MPSM.

4.4.3.1 Scenario Use Case 1: Create a Day-Ahead Action Plan

The first use case, as shown in Figure 4.13, describes a scenario to plan a day-ahead action plan in phase 1, as mentioned before in Section 4.2.

The use case *create day-ahead action plan* can at first be divided into four main activities. At first, the data needed to create the day-ahead action plan will be queried. The *request core data* services retrieve the data for all registered resources from a database. Next, it is necessary to get the information about the future action plan; therefore, it is necessary to call the *query resource planning* and the *query production demand planning* services. With the replied information the internal optimization algorithm creates the necessary schedules, and in the last step theses schedules will be distributed to the certain resources.

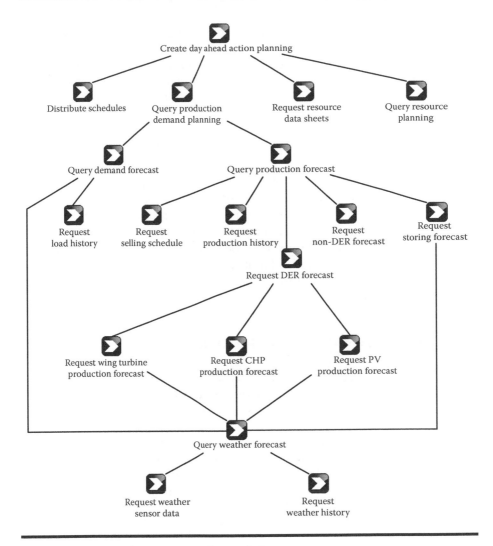

Figure 4.13   Scenario use, case 1, in graph form.

In this scenario, we elaborate the *query production demand planning* and divide it further into two services, which handle the forecast on the production and the demand on the next day.

The production forecast service on this part needs information about the already planned action on the production and storage resources. To get this data, the *request non-DER forecast*, the *request storing forecast*, as well as the *request DER forecast* service have to be called. The *request selling request schedule* provides the already placed market bids, and the *request production history* gives historical data, which additionally can be used to refine the forecasting result. An accumulated production forecast will then be generated based on these data.

The *query demand forecast* service uses the data from *request load history*, which, as the name states, delivers historical data on the demand and *query weather forecast* to generate a forecast for the demand.

The previously mentioned *request DER forecast* service is divided into specific production forecasts: the *request wind turbine production forecast*, *request CHP production forecast*, and *request PV production forecast*. Each of them estimates the production schedule for the specific production resource based on the input from the requested *query weather forecast* service.

Finally, the last services in the chain are the *request weather history* and *request weather sensor data* services, which are called from the *query weather forecast* service to reply to requests with the best possible forecast.

Regarding this use case, we are interested in internal services, which rely on external services. These may become critical in this planning scenario, where availability and reliability are crucial to provide the appropriate amount of power and to avoid instabilities. Table 4.2 describes this exemplary concern in more detail.

Table 4.2   Viewpoint—Internal Services Dependency

Name	Internal Services Dependency
Description	External and internal services can be used to operate a VPP. Moreover, internal services can rely on external services to provide their functionality, which are operated by an internal organization unit or external enterprise. This viewpoint is to depict internal services and their responsible organization unit. A distinction is being made concerning the usage of internal and external services to provide the functionality.
Intent	Using external services to render functionality reduces the influence of the VPP operator in terms of availability, reliability, and time criticality. Thus, these internal services relying on external services can be more critical than others. In the case of a failure, it is furthermore necessary to identify the organization that is responsible for maintaining the service. Moreover, this viewpoint can be used to identify potential services that should rather rely on internal services due to their criticality for the VPP. Considering these points can improve the reliability of the VPP.
Stakeholder	Data engineer, technology engineer, security engineer
Visualization	Cluster map
Required Data	Service (organization unit)
Data Source	Process engine, SLAs/contracts

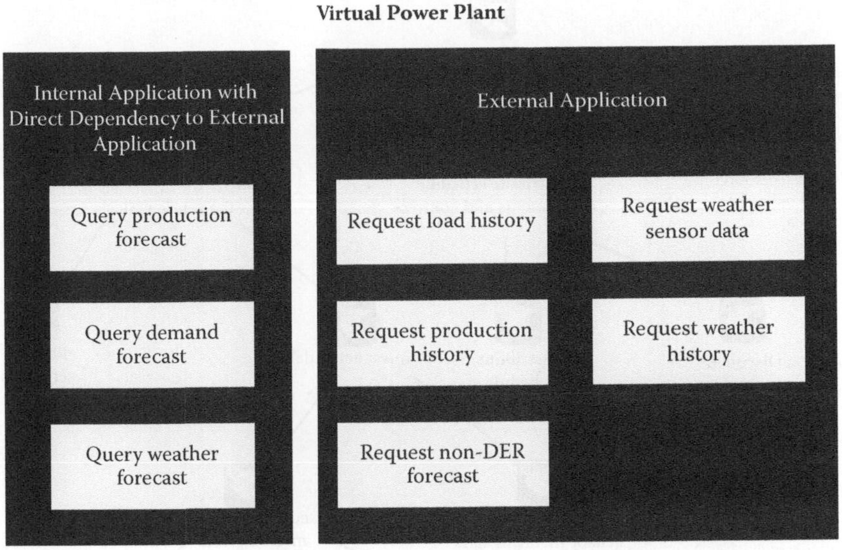

Virtual Power Plant

Internal Application with Direct Dependency to External Application	External Application	
Query production forecast	Request load history	Request weather sensor data
Query demand forecast	Request production history	Request weather history
Query weather forecast	Request non-DER forecast	

Figure 4.14   Scenario use case 1 query result.

The result of this viewpoint for our example can be seen in Figure 4.14. There are several critical services (e.g., *request load history* or *request weather history*), and some of them might especially be considered to be reorganized or to implement this functionality on the VPP site.

4.4.3.2 Scenario Use Case 2: Create a Reactive Action Plan

The creation of a reactive action plan is part of the second use case. In Section 4.2, it is mentioned as phase 2 and is visually shown in Figure 4.15.

In most cases, there will be an event or a user action that starts the process to *create a reactive action planning* to compensate eventual discrepancies between the planned and current schedules of production and consumption. Similar to the first scenario use case, the task can be divided into different activities, which can be distributed to different services. For every activity, a service has to provide the functionality, but one service can of course provide the functionality for more than one activity.

At first, the core data of the resources will be needed again, which will be done reusing the *request core data* service. The next information needed will be delivered by the service *estimate the difference to the action plan*, that will be used to determine the necessary amount of energy to shift. The core data and the estimated difference will then be used internally to optimize the action plan. The last step will include the calls of the *distribute of control signals* service to send messages like a switch-off command and the *distribute schedule* service to distribute the adjusted schedules.

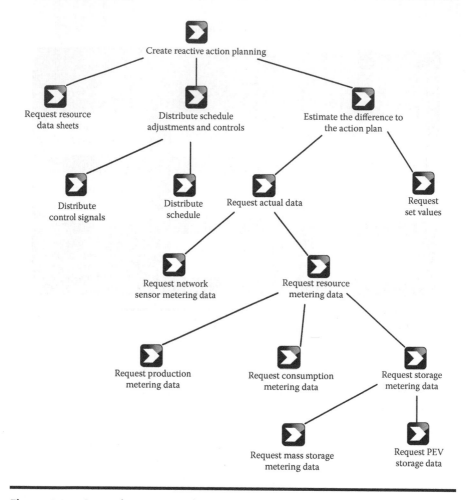

Figure 4.15 Scenario use case 2 in graph form.

To estimate the difference between the action plan and the actual situation, it is necessary to get both values. Therefore, the services *request set values* and *request actual data* are called.

The actual data have more than one source. To get an accurate view, it is required to get data about the resources themselves through the *request resource metering data* service and data about the network, which can be acquired from the *request network sensor metering data* service.

The *request resource metering data* accumulates all metered data from resources. First, there are the consumption metering data, respectively, the production metering data and, of course, the actual storage metering data, which requests on the one hand the mass storage metering data and on the other hand storage metering of small resources like PEV. An internal algorithm then calculates the entire current production and consumption.

Table 4.3   Viewpoint—OPC-UA Usage

Name	OPC-UA Usage
Question	Which VPP service can be accessed via OPC-UA or uses OPC-UA to access other services?
Description	Services can be accessed using different standards as, for example, OPC-UA or IEC 61850. The consistent usage of a defined access protocol and a corresponding data model allows services to interact with each other. All services that specifically use OPC-UA to communicate should be determined for this concern.
Intent	Using different protocols and data models requires a mapping between different services that impedes integration and comes along with potential information loss. Thus, the consistent usage improves interoperability and can allow use of standard components, preserve existing information, and finally, reduce the financial expenses needed. Due to the overview given, gaps where OPC-UA is not consistently used can be identified and also potential information loss can be identified. This analysis could further be input for security considerations. These gaps can be used to plan further developments to achieve a strategically adopted, seamless OPC-UA.
Stakeholder	Data engineer, technology engineer, security engineer
Visualization	Dependency graph
Required Data	*Service (standard)*
Data Source	Process engine

At this point, we are interested in a special issue concerning the usage of open connectivity unified architecture OPC-UA communication standard to access services, which shall be strategically adopted in our example. Rohjans et al. [26] describe a possible OPC-UA-based architecture using annotated metadata for smart grids; thus the concerned issue is a realistic one. Realizing these activities as services requires them to be integrated into the VPP to operate smoothly.

We specified a viewpoint in Table 4.3, which expresses the intention.

The result of this viewpoint is depicted in Figure 4.16. The service *request mass storage metering data* is not directly accessed using OPC-UA and is underlaid with an surrounded in shaded text, which indicates that the service itself is capable of using the OPC-UA standard, but the calling service is not. The cause is that the services *request resource metering data* and *request storage metering data* are not capable of using the standard;

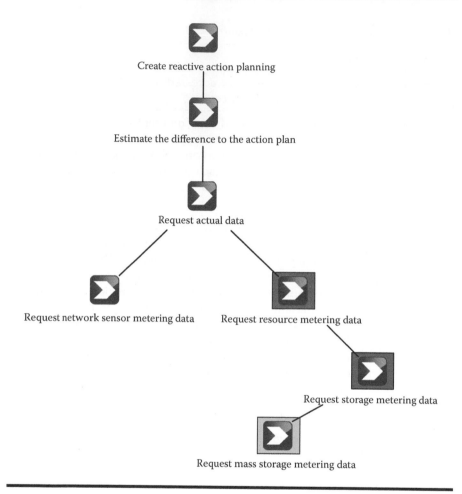

Figure 4.16   Scenario use case 2 query result.

thus a mapping is needed. The consequences of this fact can lead to a modification or a new development of this service to allow a seamless integration with the *request mass storage metering data* service.

4.5 Summary and Outlook

This chapter introduced an application of MPSM for virtual power plants. The concept of VPP was introduced in Section 4.2. VPPs may combine numerous decentralized electric generating units and consumers using a smart ICT infrastructure. From the information technology perspective as well as the utility perspective, VPPs can be realized and operated by heterogeneous and distributed services in enterprise SOA environments. A variety of stakeholders are involved in SOAs, and their concerns and views on the architecture have to be considered in a holistic way. To integrate these aspects and harmonize the different viewpoints, the MPSM approach is introduced in Section 4.3. A multiperspective SOA management approach is also imperative to successfully align both business and IT, since many enterprises are adopting or have already adopted this paradigm. Essential terms and definitions were considered in Section 4.3.1 to create a common understanding of the approach. Furthermore, the challenge of different perspectives, comprising dynamic and static information about services, business and IT views, and context-free as well as context-aware metadata, is addressed in Section 4.3.2. Section 4.3.3 introduced a metamodel and offers a template for MPSM models. Adequate information systems have to support MPSM and a schematic overview of our MPSM tool was presented in Figure 4.10. However, we refrained from describing dedicated repository technology as well as model query techniques and our visualization framework to focus on the core concepts of MPSM.

The application of the MPSM approach in VPP requires a VPP-specific structure in terms of stakeholders, viewpoints, and an adapted information model. Section 4.4 described how the automation pyramid was adopted to cover the structured EA layers. In particular, the enterprise level, operations management level, process control level, and control level were considered for architecture and service structuring. Specific attributes for each layer were defined as described in Section 4.4.2 to develop a specific information model. Finally, the MPSM-approach was exemplarily demonstrated based on a VPP use case, where two exemplary viewpoints were analyzed (Section 4.4.3).

This contribution could only show a small excerpt of VPP stakeholders, their respective concerns, and the viewpoints to be considered for VPP management. The selected use cases and services are established in the scientific community. However, the domain lacks real operation data. This problem could be solved in the near future, when more and more service-based VPPs are realized, which seems to be the most probable scenario in this domain. The results from practical experiences would provide valuable insights in management issues, and so provide input for further method development as well as the possibility to describe the specific application domain more accurately. The application area of VPPs provides a basis for further viewpoints to describe and analyze the domain. Therefore, other domain-specific layers could be chosen to cover the EA layers and also other attributes, or metadata in particular, could be annotated to the services. New technology and tools could even allow this information to be extracted from descriptions with formal semantics. The different views on VPPs can moreover use other types of visualizations, which could be linked with further information in an integrated management information system. However, it should have become clear that the MPSM method explicitly tries to consider the stakeholders' needs as the basis, and so the underlying viewpoints and models should always be adapted to the specific organization.

By now, MPSM is restricted to support documentation and analysis. Future VPP management systems might be extended by operational aspects to actively control the considered services and workflows. Beyond the application area of VPP, the MPSM approach is versatile and applicable within the energy domain whenever a service-oriented solution is chosen, for example, in the overall context of electric mobility. Especially in this area, a rapidly increasing number of services dealing with electric mobility-specific applications are expected and have to be managed.

In summary, the MPSM approach seems to be a suitable means to cope with different views on the same service landscape. As it is a generic procedure, it could be shown that it is applicable for VPP in the energy domain by exemplary use cases and analysis. Further work should consider other domain-specific application areas, additional analysis-specific visualization types, as well as more detailed information models in order to provide a better semantic basis for the analysis.

References

1. NIST. NIST Framework and Roadmap for Smart Grid Interoperability Standards, Release 1.0. Technical report, National Institute for Standards and Technology (January 2010).
2. The Open Group. TOGAF, version 9 (2009).
3. U.S. Department of Defense. The DoDAF Architecture Framework, version 2.0 (May 2009).
4. M. Postina, S. Rohjans, U. Steffens, and M. Uslar. Views on Service Oriented Architectures in the Context of Smart Grids. In *First IEEE International Conference on Smart Grid Communications*, Gaithersburg, MD, pp. 25–30 (2010).
5. K. E. Bakari and W. L. Kling. Virtual Power Plants: An Answer to Increasing Distributed Generation. In *IEEE SmartGridComm 2010*, p. 1 (2010).
6. J. Østergaard. European SmartGrids Technology Platform—Vision and Strategy for Europe's Electricity Networks of the Future (2006).
7. EREC. Renewables 24/7—Infrastructure Needed to Save the Climate (2010).
8. Agentur fuer Erneuerbare Energien. Hintergrundpapier: Das Kombikraftwerk (2008).
9. M. Stifter. IEA Demand Side Manangement—Task XVII Integration von verbraucherseitigen Maßnahmen verteilter Erzeugung, erneuerbarer Energieressourcen und Energiespeicher (2010).
10. S. Buckl, A. Ernst, J. Lankes, and F. Matthes. Enterprise Architecture Management Pattern Catalog (Version 1.0, February 2008). Technical report, Chair for Informatics 19, Technische Universität München (February 2008).

11. P. Gringel and M. Postina. I-Pattern for Gap Analysis. In G. Engels, M. Luckey, A. Pretschner, and R. Reussner, eds., *Software Engineering (Workshops)*, vol. 160, *LNI*, pp. 281–292. GI (2010).
12. The Open Group. Archimate 1.0 Specification—Technical Standard (2009).
13. IEEE 1471. IEEE Recommended Practice for Architectural Description of Software-Intensive Systems (2000).
14. M. Postina, J. Trefke, and U. Steffens. An Ea-Approach to Develop SOA Viewpoints. In *Enterprise Distributed Object Computing Conference (EDOC), 2010 14th IEEE International*, pp. 37–46 (2010). 10.1109/EDOC.2010.25.
15. A. Wittenburg. *Softwarekartographie: Modelle und Methoden zur systematischen Visualisierung von Anwendungslandschaften*. PhD thesis, Technische Universität München, Faculty of Informatics (2007).
16. T. Erl. *Service-Oriented Architecture: Concepts, Technology, and Design*. (Prentice-Hall, Upper Saddle River, NJ, 2006).
17. D. Krafzig, K. Banke, and D. Slama. *Enterprise SOA: Service-Oriented Architecture Best Practices*. The Coad series (Prentice-Hall, Upper Saddle River, NJ, 2006).
18. M. N. Josuttis. *SOA in Practice*. Software Engineering (O'Reilly, Sebastopol, CA, 2007).
19. G. Engels, A. Hess, B. Humm, O. Juwig, M. Lohmann, J.-P. Richter, M. Voß, and J. Willkomm. *Quasar Enterprise: Anwendungslandschaften serviceorientiert gestalten* (dpunkt.verlag GmbH, Heidelberg, 2008).
20. D. Steinberg, F. Budinsky, M. Paternostro, and E. Merks. *EMF: Eclipse Modeling Framework* (Addison-Wesley, Boston, MA. 2009).
21. B. Travica. The Design of the Virtual Organization: A Research Model. In *Proceedings of the Americas Conference on Information Systems*, August, pp. 15–17 (1997).
22. IEC. IEC 62264-1 ed1.0: Enterprise-Control System Integration—Part 1: Models and Terminology (2003).
23. T. J. Williams. *The Purdue Enterprise Reference Architecture* (Instrumentation Systems, Instrument Society of America, North Carolina (1992).
24. U. Katzke, *Spezifikation und Anwendung einer Modellierungssprache für die Automatisierungstechnik auf Basis der Unified Modeling Language (UML)*. (Kassel University Press, Kassel, Germany 2009).
25. L. Winkels. *Referenzmodell für die Tageseinsatzplanung dezentraler heterogener Energieerzeugungsanlagen*. PhD thesis, Universität Oldenburg (2009).
26. S. Rohjans, M. Uslar, and H. J. Appelrath. OPC UA and CIM: Semantics for the Smart Grid. In *Transmission and Distribution Conference and Exposition, 2010 IEEE PES*, pp. 1–8 (2010).

Chapter 5

Electric Distribution Grid Optimizations for Plug-in Electric Vehicles

Aline Senart, Christian Souche, Philippe Daniel, and Scott Kurth

Contents

The expected widespread adoption of plug-in electric vehicles (PEVs) is likely to challenge the electric grid. The distribution infrastructure might suffer the most, as PEV charging can substantially alter the neighborhood load and put additional stress on local equipments, such as transformers and power lines. Distribution system operators (DSOs) therefore need to prepare for the impact on the electrical system. In this chapter, we present a novel web-based planning and optimization tool that simulates without a priori assumption the technical solutions needed to support the potential widespread use of PEVs. In particular, the tool describes the potential future times and points of failures on the electric distribution grid and provides a selection of optimization programs with associated costs. DSOs can choose the most relevant programs for their network and see the benefits of applying them. Different simulations using the tool show that vehicle charging will impact the grid in the intermediate and long term.

5.1 Introduction

Most automakers are today reaching a turning point as they begin to gradually transition toward electrification of vehicles ranging from plug-in hybrids to all-electric cars [1]. With the widespread arrival of electric vehicle supply equipment (EVSE) networks, worldwide adoption of plug-in electric vehicles should grow quickly over the next 5 years [2]. By 2015, more than 3.1 million PEVs are expected to be sold worldwide [3]. This rising number of PEVs on the road could generate an additional load that would dynamically spread on the existing electricity networks both geographically and in time.

The PEV technology is considered the most influential energy technology in the grid development within the next 5 years [4]. Numerous reports have demonstrated that the planned generation capacity is likely to be sufficient for recharging vehicles, especially if the electricity demand increase occurs off-peak [5, 6]. However, the negative impact on the electrical distribution networks (EDNs) that transport electricity to the end user could be substantial. Studies have shown that PEV charging could be clustered in specific neighborhoods [7]. Therefore, even if the PEV penetration remains globally low, adoption in local areas could become very high. Depending on the PEV penetration level, charging requirements, timing, and duration of the vehicles' connection to the grid, a local EDN could become overloaded and the life span of power equipment, such as switching equipments, transformers, and regulators, could be shortened.

It is therefore important for grid operators to anticipate the potential impact that PEVs could have on the electric distribution. Good planning and optimization tools helping them to assess the impact and invest into EDN optimizations are necessary. Previous work has primarily focused on determining well-to-wheel vehicle emissions [5, 8], power supply adequacy [5], or the benefits of vehicle-to-grid [9, 10]. Only few studies have addressed the impact assessment of PEV load on the grid infrastructure at the local level. Most of them are research methodologies based on surveys [11, 12] or relatively small-scale network simulations focused on determining the threshold of PEV penetration supported by a class of distribution assets [13, 14].

In this chapter, we present a strategic planning and optimization tool that can be used to inform decisions about whether and how to optimize the distribution grid in order to support the increasing number of PEVs. The proposed tool doesn't rely on any a priori assumption and enables distribution grid operators to easily analyze alternative market scenarios by simulating efficiently and accurately different EDNs with different parameters. In a first step, the tool forecasts the future problems on the grid and their timeframe. The tool further suggests programs for optimizing the EDNs to cap this impact. For example, the tool might suggest where reinforcement of the grid may be necessary or what incentives to put in place to encourage consumers to plug in their PEVs at the right time. Finally, the DSOs can select the appropriate programs and visualize the subsequent improvements of applying the optimizing measures to the EDNs. Costs and benefits associated to the programs help utilities prioritize their investments.

An evaluation of the impact of PEV charging on different EDNs is carried out. Research findings indicate that in most networks PEV charging will create load peaks that strain equipment beyond its designed specifications. This will result in a decrease in the operating efficiency of the distribution network and will require network upgrades or higher control in charging patterns.

This chapter is structured as follows. Section 5.2 reviews the state of the art on PEV impact estimations. Section 5.3 describes the proposed tool to help DSOs plan for PEV market uptake. Section 5.4 presents results obtained from using our tool in various scenarios. Finally, Section 5.5 provides conclusive remarks and presents future work.

5.2 State of the Art

There have been many publications around plug-in electric vehicles: from estimating their oil consumption, emissions, costs, consumer adoption, power supply adequacy, to the potential of vehicle-to-grid. We focus in this section only on existing studies that have attempted to assess their impact on the distribution electric system.

5.2.1 Research Methodologies

To date, most impact studies have developed methodologies or spreadsheet models that can be used by electrical utilities to analyze the effect of PEV charging on the distribution system under various scenarios. The models have been developed based on interviews, surveys, U.S. census data, tax assessment data, and other sources [15].

An example of research methodology is presented in [11]. The authors have developed a spreadsheet model to assess the impact of PEV adoption on the entire San Francisco County between 2010 and 2020. They also provide a list of issues and recommendations for accelerating PEV adoption and charging infrastructure deployment in the Bay Area.

Another research methodology is presented in [12], where near-term recommendations are provided to manage the impacts of PEVs on the electric grid in Dane County, Wisconsin. The impact assessment relies on three aspects: infrastructure readiness assessment, consumer preference analysis, and grid impact studies.

Finally, [16] is a PEV readiness study that identifies which cities in America are currently ready for PEVs, including an analysis of the necessary infrastructure investments to prepare for local adoption. The study is based on a combination of research and city stakeholder interviews conducted from January through August 2010.

Even though research methodologies and models are useful to understand the impact in specific areas, they are difficult to apply in an automated way to other geographies and distribution networks. Moreover, the assessment is coarse-grained and doesn't include the impact on specific network assets.

5.2.2 Distribution Network Simulations

Some studies have therefore developed impact models that are implemented over a network simulation platform. This enables one to easily simulate the behavior of electrical distribution networks under different operating conditions and scenarios.

These studies are, however, limited in scope. In [14], a subset of feeders is analyzed. A statistical clustering algorithm is used to identify a set of representative feeders for the electric utility system and the impact analysis is only performed on them. The PDCIM model defined in [17] is also used to only estimate the impact on transformers and underground cables and not the impact on the whole distribution system.

Moreover, most impact models do not offer enough flexibility to accommodate any scenario. For example, the analytical framework developed in [18] is intended to evaluate the impacts of PEVs on distribution system thermal loading, voltage regulation, transformer loss of life, unbalance, losses, and harmonic distortion levels. Strong assumptions about plug-in hybrid vehicle penetration have, however, been taken to generate projections on the number of vehicles per utility customer. The authors also neglect EVs in their study to concentrate on plug-in hybrid vehicles only, as they believe that the market share for these vehicles will be small.

Another major limitation of existing systems is their simplicity. The models do not take into account the many factors affecting the PEV impact. For example, in [17, 18, 19], the focus is on home charging, and PEV charging at work or in public places is not considered. The load is often randomly distributed to residential customers, with a maximum of two vehicles per customer, as seen in [17]. Finally, the scales of the simulations are relatively small [20], and the robustness of the systems can be questioned.

Previous studies based on simulations are promising, but unfortunately none of this work provides an effective planning and optimization tool that could be used by DSOs to evaluate the impact of PEVs on their distribution grid and possible solutions to cap this impact.

5.3 Electric Distribution Grid Optimizations

At Accenture Technology Labs we developed a generic web-based tool that can be used by DSOs to run different scenarios with varying PEV market penetrations for different distribution networks.

5.3.1 System Capabilities

One of the challenges for such a tool is that the number of plug-in electric vehicles sold, the concentration of their usage, and the rapidity of the market growth are unknown. Rather than imposing arbitrary restrictions, we provide a flexible tool that allows the user to specify PEV assumptions, the geographical environment (be it the U.S., Europe, a city, a suburban area, etc.), and the conditions (without fast charging, without charging incentives, etc.) under which the simulations will be performed.

The tool provides visualization capabilities to locate potential future stress on the distribution grid or network deficiencies with their time of occurrence. A personalized selection of optimization programs that address the problems identified or improve the operation of the EDNs is further offered to the grid operator. Such optimization programs include:

1. Making the necessary upgrades of the EDN assets to support the additional load, e.g., prioritization of asset investments for transformer and

conductor upgrades when the assets are operating over their capacity or adding capacitor banks when there is voltage loss.

2. Reducing the impact on the EDNs by changing driver behavior, e.g., enforcing different charging controls (day/night tariff, real-time pricing, or price schedule) to encourage drivers to charge their PEVs during off-peak hours when the demand for energy is low.

3. Advising new EVSE locations, ranging from places of employment as well as residence or on-street charging stations to satisfy the future fleet of PEVs at a rate to match the market growth.

Some optimization programs are also directed to helping utilities plan for vehicle-to-grid (V2G) and maximize their benefits while retaining enough energy in PEVs for driving needs. Utilizing V2G may allow the EDNs to draw energy stored in the PEVs as necessary, allowing for a more efficient and flexible use of electrical energy. These optimization programs are not presented in this chapter due to space limitations.

Each of these programs is presented to the DSOs with an associated cost. The cost of a program will vary depending on the level of upgrades on the infrastructure selected and whether there is existing smart grid technologies used in the EDNs. On choosing the appropriate programs, DSOs can further make an informed decision by visualizing the improvements obtained on the distribution networks.

The advantages of using the web-based planning and optimization tool are threefold:

1. With improved visibility over which assets are loaded below or over their ratings, DSOs can better plan and prioritize physical asset/EVSE purchases and replacements and optimize future investments.

2. Through a better operational understanding of asset health and performance, DSOs can keep their load below their ratings and by consequence extend their service life.

3. A better planning and control of the load reduces the chance of unplanned outages and maintenance to grid assets, thus improving customer satisfaction and reducing the costs associated with unscheduled repair.

The following sections will present the architecture of the solution in terms of components, their relationships, and input for these different components.

5.3.2 System Architecture

The architecture of the tool is composed of three main parts, as shown in Figure 5.1. First, the front end acts as an interface for users to interact at a high level with the system. Second, the back end is an analytics engine that calculates the PEV load for each EDN asset and recommends one or more

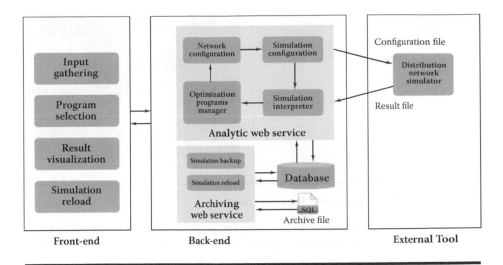

Figure 5.1   System architecture.

EDN optimizations to the DSOs. Finally, the system architecture comprises an external tool that computes the power flow in the distribution networks.

5.3.2.1 Front End

The front end is a website that has been implemented in Flex 3. The interactions with the users are through screens. Communications between the front end and the back end are via .Net web services.

The front end comprises the following components:

Input gathering: The front end allows the EDN operators to specify a set of parameters that will be used to simulate a specific EDN. These parameters are set through screens (see Figure 5.2a) and include simulation configuration details, PEV estimations, the distribution network to simulate, and the geographic area where the EDN is located (see Table 5.1).

Result visualization: The front end also displays a dynamic view of the distribution network over the years of simulation with a highlight on the potential failure points (see Figure 5.2b). Some valuable statistical results about a set of simulations that have been performed are also shown.

Program selection: Finally, the front end allows the user to apply one or many of the optimization programs established to help the EDNs handle their new load. The result visualization component can display the EDN health before and after application of the program measures.

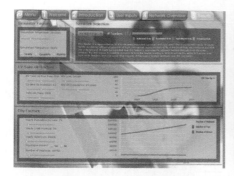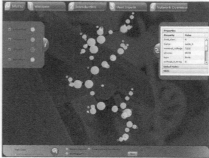

Figure 5.2 **(a) User input and (b) result pages.**

Simulation reload: Finally the user can save the current simulation and reload previous simulations through the front end. When starting a simulation, the user can decide to create a new one or to launch a previously saved simulation.

5.3.2.2 Back End

The back end has been developed with the .Net framework. Interactions with the front end and the database are performed through .Net web services, and interactions with the simulator are based on files. The database used is Microsoft SQL Server.

The back end runs k simulations to cover the time span requested by the user where k is defined as the following:

$$k = \frac{EndDate - StartDate}{Frequency}$$

At first, the simulation parameters are set, either from a previously saved simulation, or from the front end. In a first round, assessment of the baseline load profile is computed for all the EDN assets of the simulation. The PEV demand profiles are then assessed for each time step and added to the baseline load profile to obtain an accurate estimation of the total load per asset. The back end sends alerts to the front end when the set of simulations are finished so that results may be displayed to the EDN operators with possible suggestions of optimization programs. A new round of simulations are performed when the user decides to apply the programs, and the new results are displayed through the front end. The user can save the current simulation at any time.

The back end delivers simulation analytics and simulation archiving services through the following components:

Network configuration manager: This component is responsible for the distribution network modeling. It first parses the description

Table 5.1 Nonexhaustive List of Inputs

Category	Input	Description
Simulation	StartDate	Year when the simulation starts (e.g., 2010)
	EndDate	Year when the simulation ends (e.g., 2060)
	Frequency	Simulation frequency Frequency ϵ [yearly, quarterly, monthly]
PEV	PEVFinalPenetrationRate	The PEV penetration rate estimated at EndDate (e.g., 50% in 2060)
	PEVEvolutionRate	The estimated acceleration of the PEV penetration
	PEVLoad	Average energy consumed to charge a PEV (e.g., 24 kWh)
	LifeExpectancy	Average life expectancy of a vehicle (electric or not) in years (e.g., 13)
	NumberCarSales	The total number of vehicles sold at StartDate
Distribution network	N	The total number of assets
	InitialLoad$_i$	The initial load on asset i at StartDate
Geographic area	TotalPopulation	The total population of the geographical area
	YearlyPopulationIncrease	The estimated percentage of increase of the population every year (e.g., 0.02)

of the EDN to be used for a set of simulations and translates the description into our network data model. From the input gathered from the front end, the network manager also computes the baseline load and the load corresponding to each simulation date and puts that load in the corresponding tables of the database. The network manager uses the load model discussed in Section 5.3.3.1 to compute this load. Finally, the network manager can be used to define a new configuration of the EDN. The new configuration is computed according to the optimization programs that have been selected.

Simulation configuration manager: This component configures the different simulations to run, specifying their different parameters. Before each simulation, it generates from the database a script file describing the configuration of the EDN for the simulation date, and passes it to the simulator. The simulator in turn runs the simulation to simulate the EDN operation for a specified period of time according to the script file and generates result files.

Simulation interpreter: This component parses and interprets the result files produced by the simulator. The result files may include information on various aspects of the EDN, such as currents, voltages, and power values of the assets in the EDN. The interpretation may detect various future issues within the EDN for the load that has been applied, such as, for example, outages and line failures.

Optimization programs manager: After all of the simulations are complete and the results are interpreted, the optimization programs manager may determine, based on the results, one or more improvement programs that may address the problems identified or improve the operation of the EDN. If selected by the EDN operators, the component may iterate through the process. For example, it may ask the network manager to update the EDN configuration to reflect the optimized EDN, and run the simulation again with this optimized EDN configuration. Once the new results are interpreted, the optimization programs manager informs the front end to display results to the EDN operators.

Database: All information gathered as input, intermediate results, and final interpretation of the simulations are stored in a database. A network data model of a classical EDN is defined in the database. This model includes database tables containing the characteristics of each asset of the EDN and the relationships between the assets forming a grid topology. Receiving the simulation data from the EDN operators as user input and storing in the database allows the system to dynamically simulate any EDN with varying configurations.

Simulation backup: The user can save the current simulation parameters at any time during the simulation in an archive file. All the needed information for efficient recovery is automatically selected and saved. This will enable the user to easily and quickly run the same or similar simulation in the future.

Simulation reload: Saved simulation can be restored by the user at any time. The restored simulation can be edited to reflect new infrastructural changes in the network, or reused directly to perform additional network assessments or to test new optimization programs.

5.3.2.3 External Tool

Simulations have been performed based on the EDN configuration information using an off-the-shelf simulator called GridLAB-D™[21]. GridLAB-D is a power distribution system simulation and analysis tool, developed by the U.S. Department of Energy (DOE) at Pacific Northwest National Laboratory. The simulator receives the description of an EDN as a file. The simulator simulates the EDN for a specific range of time and generates result files containing electrical values such as currents and voltages. Those results files are parsed by the simulation interpreter to populate the database and will be later used to suggest optimization programs. Note that the power flow simulation could have been provided by any algorithm implementing the three-phase forward-back sweep method.

5.3.3 System Optimization

This section presents the analytics that have been developed to infer the optimization programs for a particular network.

5.3.3.1 Electric Load Model

A load model is used by the network manager to compute the load corresponding to each simulation date. The load on a network asset is composed of the base load plus the PEV load. The current base load is known, but the population growth and the increase in individual power needs have to be taken into account for future base loads. The current PEV load, however, is unknown and accounts for factors such as, for example, energy consumption of a PEV, the charging profile of the batteries used, and the driving behavior of the drivers.

The load at each node and for each simulation may be computed with a set of formulae. These formulae assume that the total demography and the base load for each node are known at the starting date. In those formulae, n represents the date of simulation and i represents a loaded node.

The first parameter to compute is demographyweight. It represents the weight of each node as far as demography is concerned. The demography of a node is the number of people that receive their electric power from that node. The formula also uses the YearlyPopulationIncrease corresponding to the simulation date. The last element used by formula is ActivityInfluenceonDemography. It is a coefficient between 0 and 1, and represents the influence that the activity on a region has on its demography. For example, four regions may be defined as commercial, residential, agricultural, and industrial.

$$\text{DemographyWeight, } (n) = \frac{Load_i(StartDate)}{\sum_{i=n}^{N} Load_i(StartDate)}$$
$$\cdot(1 + YearlyPopulationIncrease)^n \qquad (5.1)$$
$$\cdot\text{ActivityInfluenceonDemography}(i)$$

Given the previous parameter, the overall demography at the start date and the YearlyPopulationIncrease demography at each node may be computed according to Equation 5.2.

$$\text{Demography}_i(n) = \frac{\text{DemographyWeight}_i(n)}{\sum_{i=n}^{N} \text{DemographyWeight}_i(n)}$$
$$\cdot\text{Demography}(StartDate) \qquad (5.2)$$
$$\cdot(1 + YearlyPopulationIncrease)^n$$

PEVWeight, calculated by Equation 5.3, represents the likelihood of each node to handle some PEVs. This is computed using the demography at each loaded node and ActivityInfluenceonPEV, which is a coefficient between 0 and 1, representing the influence the activity in a region has on the number of PEVs used within a region.

$$\text{PEVWeight}_i(n) = \frac{\text{Demography}_i(n)}{\sum_{i=0}^{N} \text{Demography}_i(n)}$$
$$\cdot\text{ActivityInfluenceonPEV}(i) \qquad (5.3)$$

Equation 5.4 computes the number of PEVs that will be recycled during the year corresponding to the simulation date. It uses the life expectancy of each PEV and the number of cars sold for number of years corresponding to the life expectancy before the year of simulation.

$$\text{PEVToRecycle}(n) = \text{ if}((n - lifeExpectancy) >= StartDate)$$
$$\text{then NumberCarSales}(n - lifeExpectancy) \text{ else } 0$$
$$where \text{ NumberCarSales}(n) = \text{NumberCarSales}(StartDate)$$
$$\cdot(1 + YearlyPopulationIncrease)^n \qquad (5.4)$$

Equation 5.5 computes the number of PEVs for the current simulation date. That equation uses the number of PEVs of the previous simulation, the number of PEVs to recycle, the PEV penetration, the number of cars sold at the start date, and the corresponding YearlyPopulationIncrease.

$$PEVPenetration(n) = \frac{PEVFinalPenetrationRate}{1 + e^{((InflexionDate-n) \times PEVEvolutionRate)}} \quad (5.5)$$

$$with \ n \in [StartDate, EndDate]$$

The parameter NumberPEV(n) used with PEVWeight at each node helps compute the number of PEVs at each node for the simulation date according to Equation 5.6.

$$NumberPEV_i(n) = \frac{PEVWeight_i(n)}{\sum_{i=0}^{N} PEVWeight_i(n)} \cdot NumberPEV(n) \quad (5.6)$$

Using Equations 5.1 to 5.6, all the parameters to compute the load at each node according to Equation 5.7 are obtained. PEVLoad represents the average power used by a PEV:

$$Load_i(n) = Load_i(StartDate) \cdot (1 + YearlyLoadIncrease)^n$$

$$\cdot (1 + YearlyPopulationIncrease)^n \quad (5.7)$$

$$+ NumberPEV_i(n) \cdot PEVLoad$$

5.3.3.2 Simulation Interpretation

After running a simulation, the simulator provides the results by generating result files. The simulation interpreter parses the generated result files and stores result values at the corresponding places in the database. The results that have been input to the database are then interpreted to determine if the EDN worked properly or not for the load that has been applied.

The simulation interpreter determines whether the network assets are aging, detects any overcapacity situations, detects any faults on recloser/fuse, and detects any voltage drops. For example, by comparing the estimated power and the nameplate power rating of transformers, the simulation interpreter infers which could be working at their edges and are going to age more quickly. For example, transformers above 150% of their rating are overloaded and will see their expected life reduced. Upgrades therefore need to be planned (e.g., replacing a 160 kVA transformer by a 250 kVA transformer). Similarly, when a fault occurs on the distribution system, it is interrupted and cleared by a fuse, recloser, or relayed circuit breaker. The simulation interpreter compares the estimated current to the equipment design specifications to anticipate a potential fault on a line. Aging is estimated from the IEEE Standard C57.91-1995 load-dependent failure rate given the load on the transformer and an ambient temperature of 25°C [22].

Finally, power losses and voltage drops are determined by comparing current power or voltage to historical values in the database. According to European Norm EN50160, voltage deviations should be less than 10% for 95% of the time.

5.3.3.3 Program Selection

After all of the simulations are complete and the results are interpreted, the back end may determine, based on the results, one or more improvement programs that may address the problems identified or improve the operation of the EDN. The one or more improvement programs may be directed to one or more goals, such as minimizing power loss, voltage dip, and avoiding asset overload within the EDN.

The optimization programs manager generates a plan for automatically implementing the above improvement programs, and may be operable to communicate with the EDN via the network to automatically modify the EDN assets according to the plan. The optimization programs manager may also iteratively run new simulations taking into account the improvement plan, generate further improvement plans based on the new simulations, and make necessary changes to the EDN according to the new improvement plans. The costs of the programs include equipment purchase, labor cost to remove and install the equipment, software updates, and smart grid development when appropriate.

5.4 Results

We focus in this section on the insights obtained by the first optimization program. Please note that only worst-case scenarios at peak demand are evaluated.

5.4.1 Assets Condition

To evaluate the future conditions of distribution network assets, we ran the planning and optimization tool on different networks under different scenarios over 20 years. The simulations are performed with two different goals.

5.4.1.1 Assets Condition on Different Networks

For this evaluation, the yearly population increase is set to 0.5 and the yearly load increase to 1.9. The PEV take-up scenario is optimistic with an inflection date for the PEV take-up in 2021 and the PEV penetration rate at 50% in 2031.

The networks are the following:

- **Network 1:** This feeder is a representation of a lightly populated rural area. The load is composed on single-family residences with some light commercial. Approximately 88% of the circuit-feet are overhead and 12% underground, 100,000 inhabitants, 230 nodes, 125 residential and commercial transformers.
- **Network 2:** This feeder is a representation of a moderately populated suburban and lightly populated rural area. This is composed mainly of single-family residences with small amounts of light commercial. Approximately 70% of the circuit-feet are overhead and 30% underground, 500,000 inhabitants, 337 nodes, 230 residential and commercial transformers.
- **Network 3:** This is composed mainly of single-family residences with small amounts of light commercial. Approximately 60% of the circuit-feet are overhead and 40% are underground, 1,000,000 inhabitants, 613 nodes, 452 residential, commercial, and agricultural transformers.

The planning and optimization tool has been used with much bigger networks, but running the simulations can take time (sometimes several hours). While this is not an issue when the DSO goal is to predict long-term PEV impact, it is hampering when thousands of simulations have to be launched to evaluate the tool.

Figure 5.3 displays the increasing number of PEVs computed for these distribution networks according to Equation 5.6 presented above. As anticipated, the number of PEVs is directly proportional to the size of the networks. As the network size increases, the number of inhabitants increases and so does the number of PEVs.

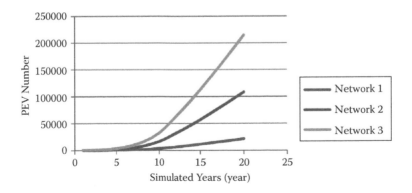

Figure 5.3   Number of PEVs per year.

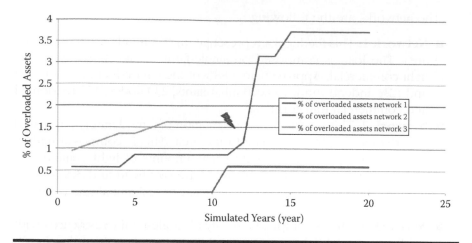

Figure 5.4   Percentage of overloaded assets by year.

To get an estimate of the PEV impact on network health, we plotted in Figure 5.4 the percentage of overloaded assets per year for the three networks.

Several interesting facts can be observed in the figure. First, in some cases, few assets are overloaded already in year 1. These assets are those that are currently running close to their nameplate specifications. With only few PEVs deployed, they become overloaded. Second, a threshold in year 10 (corresponding to 2021) can be noticed. By looking at Figures 5.3 and 5.4 together, this year corresponds to the uptake of PEVs with our simulation inputs. Third, the maximum percentage of overloaded assets is capped at 3.8% given our assumptions. The impact is therefore relatively small. However, network 3 experiences an outage in mid-2021. Strategic assets being overcapacity for too long might bring the complete network down. Our tool will therefore become essential to operators to make asset replacement decisions and help prevent future outages.

5.4.1.2 Assets Condition with Different Scenarios

For this evaluation, we compare the effect of different PEV penetrations on network 2. The assumptions for the simulations are the following. The PEV final penetration rates are respectively set to 15, 30, and 50% for scenarios 1, 2, and 3. The inflection date in 2021 and the evolution rate of 0.5 are, however, shared for the three scenarios. Figure 5.5 shows the percentage of overloaded assets by year for the different scenarios.

Results show that the level of PEV penetration has a direct impact on the percentage of overloaded assets. The number of assets used over their capacity increases with the number of PEVs in the streets.

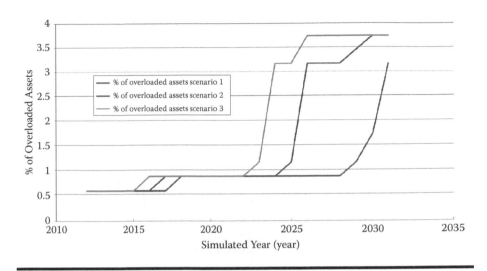

Figure 5.5   Percentage of overloaded assets by year.

5.4.2 Load Increase

To help us understand the causes of this impact, we plotted the load in-crease over 20 years for the three networks in Figure 5.6. The yearly PEV number computed for Figure 5.3 is used as a base to estimate the extra load applied at each node. The load displayed in the figure is an average of all the loads applied at all the nodes of the distribution network.

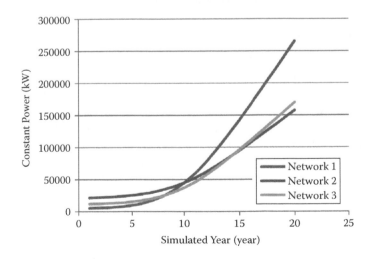

Figure 5.6   Load increase (kW).

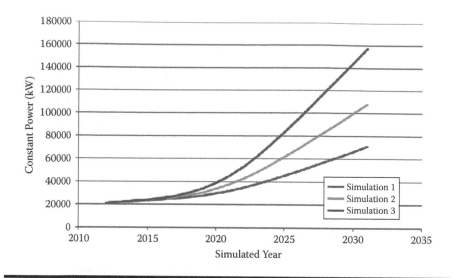

Figure 5.7   Load increase (kW).

Results show a significant increase in the electric load over the years. This increase is due to the natural population increase and natural load increase (as electric appliances are more and more energy greedy) and the new PEV load. A clear inflection can be seen in year 10 (corresponding to 2021), suggesting that the most influential factor in the load increase is the PEV load.

Similarly, Figure 5.7 demonstrates that when the PEV take-up scenarios become more optimistic, the load increases.

5.4.3 EDN Health

Distribution network health can be analyzed in terms of power losses and voltage deviations. Figure 5.8 shows the total power losses by PEV on network 2. The increase of the number of PEVs indicates a significant increase in power losses.

Not only power losses but also voltage deviations of the grid voltage are essential to the distribution grid operator as well as to grid customers. Even though there are voltage drops (as seen in Figure 5.9), the deviations are less than 10% and remain acceptable according to the European Norm EN50160.

5.4.4 Transformer Conditions

We saw in Section 5.4.1 that some assets, including transformers, get overutilized as the PEV deployment spreads. Figure 5.10 shows additional

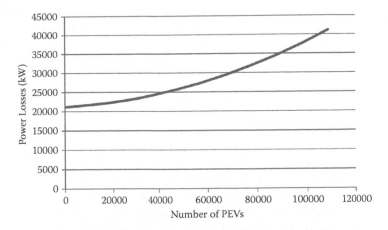

Figure 5.8   Power losses by PEV.

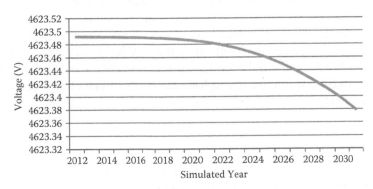

Figure 5.9   Average voltage.

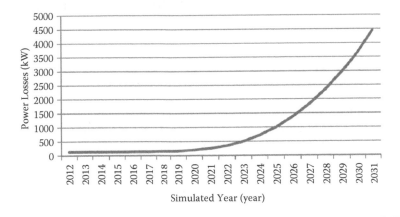

Figure 5.10   Transformer power losses.

insight: the power losses of transformers. These losses increase gradually with the years, showing that appropriate procurement and sizing of transformers specific to demand in certain networks will have to be planned.

5.5 Conclusions and Future Work

While PEVs represent a significant potential growth in energy demand, the expected influence of PEVs on the electrical distribution system has not been completely evaluated. In Accenture Technology Labs, we developed a web-based planning and optimization tool to provide the ability to our utility clients to account for any additional stresses to their systems. We conducted a detailed evaluation of the potential impacts through different scenarios. Results indicate that the deployment of PEVs will have diverse effects on the infrastructure. We discovered that most distribution networks will experience problems. The impact will remain relatively small under our simulation conditions but will, however, require maintenance and upgrade of network equipment. As future work, we are improving the electric load model to reflect the varying location of the PEV load over time, depending on working hours, seasonal changes, popular events, etc. Further research also needs to be conducted to explore the effect of fast charging compared to normal charging.

References

1. Accenture, Betting on Science: Disruptive Technologies in Transport Fuels, Accenture, Research Report, November 2009. [Online]. http://newsroom.accenture.com/article_display.cfm?article_id=4899
2. Gartner, Hype Cycle for Smart Grid Technologies 2010, Research Report 2010.
3. Pike Research, Electric Vehicles: 10 Predictions for 2010, Research Report 2010.
4. Gartner, User Survey Analysis: Energy and Utilities Industry, North America, 2010, 2010.
5. Stanton Hadley and Alexandra Tvsetkova, Potential Impacts of Plug-in Hybrid Electric Vehicles on Regional Power Generation, *The Electricity Journal*, pp. 56–68, 2009.
6. Christopher Yang and Ryan McCarthy, Electricity Grid: Impacts of Plug-in Electric Vehicle Charging, *Environmental Management*, pp. 16–20, 2009.
7. Pedram, Mohseni and Richard Stevie, Electric Vehicles: Holy Grail or Fool's Gold, in *IEEE Power & Energy Society General Meeting*, Calgary, AB, 2009.
8. Ulrich Eberle and Rittmar von Helmolt, Sustainable Transportation Based on Electric Vehicle Concepts: A Brief Overview, *Energy & Environmental Science*, pp. 689–699, 2010.

9. Mohamed El Chehaly, Omar Saadeh, Carlos Martinez, and Geza Joos, Advantages and Applications of Vehicle to Grid Mode of Operation in Plug-in Hybrid Electric Vehicles, in *IEEE Electrical Power & Energy Conference*, Montreal, QC, 2009, pp. 1–6.

10. Koichiro Shimizu, Taisuke Masuta, Yuyaka Ota, and Akihiko Yokoyama, Load Frequency Control in Power System Using Vehicle-to-Grid System Considering the Customer Convenience of Electric Vehicles, in *International Conference on Power System Technology*, Hangzhou, China, 2010, pp. 1–8.

11. Dipti Desai, Plug-in Hybrid Electric Vehicle Adoption Estimation and System Load Impact in San Francisco County: 2010–2020, San Fransisco, CA, 2009.

12. Jessica Y. Guo et al., Consumer Adoption and Grid Impact Models for Plug-in Hybrid Electric Vehicles in Dane County, Wisconsin, Research Report 2010.

13. Shengnan Shao, Manisa Pipattanasomporn, and Saifur Rahman, Challenges of PHEV Penetration to the Residential Distribution Network, in *IEEE Power & Energy Society General Meeting*, Calgary, AB, 2009.

14. Luther Dow, Mike Marshall, Le Xu, Julio Romero Agüero, and H. Lee Willis, A Novel Approach for Evaluating the Impact of Electric Vehicles on the Power Distribution System, in *IEEE Power & Energy General Meeting Panel Session*, Minneapolis, MN, 2010.

15. Kevin James Dyke, Nigel Schofield, and Mike Barnes, Analysis of Electric Vehicles on Utility Networks, in *The 24th International Battery, Hybrid, Fuel Cell Electric Vehicle Symposium*, Stavanger, Norway, 2009.

16. Antonio Benecchi, Matt Mattila, and Shamsuddin Nauman Syed, PEV Readiness Study, 2010.

17. Chris Farmer, Paul Hines, Jonathan Dowds, and Seth Blumsack, Modeling the Impact of Increasing PHEV Loads on the Distribution Infrastructure, in *43rd Hawaii International Conference on System Sciences*, Honolulu, HI , 2010.

18. Jason Taylor, Arindam Maitra, Mark Alexander, Daniel Brooks, and Mark Duvall, Evaluation of the Impact of Plug-in Electric Vehicle Loading on Distribution System Operations, in *IEEE Power & Energy Society General Meeting*, Calgary, AB, 2009, pp. 1–6.

19. Kristien Clement-Nyns, Edwin Haesen, and Johan Driesen, Analysis of the Impact of Plug-in Hybrid Electric Vehicles on Residential Distribution Grids by Quadratic and Dynamic Programming, in *The 24th International Battery, Hybrid, Fuel Cell Electric Vehicle Symposium*, Stavanger, Norway, 2009.

20. Kristien Clement-Nyns, Edwin Haesen, and Johan Driesen, The Impact of Charging Plug-in Hybrid Electric Vehicles on the Distribution Grid, in *Fourth IEEE Young Researchers Symposium in Electrical Power Engineering*, Eindhoven, The Netherlands, 2008.

21. GridLAB-D Simulation Software. [Online]. http://www.gridlabd.org/

22. IEEE Standards Association, C57.91-1995—IEEE Guide for Loading Mineral-Oil-Immersed Transformers, 1995.

Chapter 6

Ontology-Based Resource Description and Discovery Framework for Low-Carbon Grid Networks

K.-K. Nguyen, A. Daouadji, M. Lemay, and M. Cheriet

Contents

Using smart grids to build low-carbon networks is one of the most challenging topics in the ICT industry. The GreenStar Network is the first worldwide initiative completely powered by renewable energy sources across Canada. Smart grid techniques are deployed to migrate data centers across network nodes according to energy source availabilities, thus reducing CO_2 emissions to minimal. Such flexibility requires a scalable resource management support, achieved by virtualization enabling the sharing, aggregation, and dynamic configuration of a large variety of resources. Such a virtualized management is based on an efficient resource description and discovery framework, dealing with a large number of heterogeneous elements and the diversity of architectures and protocols. In this chapter, we present an ontology-based resource description framework, developed particularly for ICT energy management, where the focus is on energy-related semantics of resources and their properties. We propose then a scalable resource discovery method in large and dynamic collections of ICT resources, based on semantics similarity inside a federated index using a Bayesian belief network. The proposed framework allows users to determine the cleanest resources to fulfill their requirements, regarding the energy source availabilities. Experimental results are shown to compare the proposed method with a traditional one in terms of GHG emission reductions.

6.1 Introduction

Nowadays, reducing greenhouse gas (GHG) emissions is becoming one of the most challenging research topics in information and communication technologies (ICT) because of the alarming growth of indirect GHG emissions resulting from the overwhelming utilization of ICT electrical devices. The current approach when dealing with the ICT GHG problem is improving energy efficiency, which aims to reduce energy consumption at the micro level. Research projects following this direction have focused on microprocessor design, computer design, power-on-demand architectures, and virtual machine consolidation techniques. However, a microlevel energy efficiency approach will likely lead to an overall increase in energy consumption due to the Khazzoom-Brookes postulate (also known as Jevon's paradox), which states that *"energy efficiency improvements that, on the broadest considerations, are economically justified at the micro level, lead to higher levels of energy consumption at the macro level"* [1]. Therefore, we believe that reducing GHG emissions at the macro level is a more appropriate solution. Large ICT companies, like Microsoft, which consumes up to 27 megawatts of energy at any given time [15], have built their data centers near green power sources. Unfortunately, many computing centers are not so close to green energy sources. Thus, green energy distributed networks using smart grid technologies are an emerging technology. An important assumption to make is that losses incurred in energy transmission over power utility infrastructures are much higher than those caused by data transmission, which makes relocating a data center near a renewable energy source a more efficient solution than trying to bring the energy to an existing location.

Such a new green ICT network-based approach has to face a number of issues, such as the complexity of networks, the large number of elements, the diversity in architectures and organizations, and highly flexible requirements. The GreenStar Network [2] is the first nationwide network in the world that is powered only by green energy. Virtualization, a new paradigm being explored by the research and education community dealing with highly complex distributed environments, is deployed to address resource management issues in such large-scale smart grid networks. Virtualized management will likely become one of the key solutions interconnecting appliances in order to give each of the virtual entities a complete semblance of their counterparts. Main characteristics of a virtualization technique include: (1) a warping of network elements, such as connectivity resources and traffic processing resources, (2) dynamic establishment capability, such as flexible and efficient mechanisms to trigger and tear down service, (3) end-to-end across multiple domains, and (4) control by the end user, e.g., the end user should be able to operate the virtual infrastructure as if it were a dedicated physical infrastructure.

Key techniques supporting a scalable network management include resource description and discovery. An inefficient resource description solution could be a barrier for developing powerful resource discovery mechanisms, which in turn makes the resource management inefficient. For example, resources may have different descriptive information, using different formats. They may also be indexed using different keywords referring to various characteristics. This imposes several difficulties for searching and discovery.

In the context of virtualized resource management, this chapter presents a framework for resource description and discovery, which targets low-carbon network management based on smart grid technologies. The resource description is based on an energy-oriented ontology proposed for ICT resources. Resource information is represented using Resource Description Framework (RDF) graph models [8] and web semantics, which enables resource classification, indexing, keyword searching, and semantic analyses. The resource discovery is based on a Bayesian semantic approach to enable various resource description methods. Thesaurus-based searching is also implemented addressing the keyword uncertainty problem, where we define a priori probability on each keyword, in order to improve the performance of the resource discovery engine.

The main contribution of this chapter includes an energy-oriented ontology for ICT equipment, a description of grid resources based on proposed ontology, and a resource discovery method that takes into account the energy consumption and GHG emissions of ICT resources.

The remainder of the chapter is organized as follows. In the next section, we present the key elements to build a low-carbon network with a resource management framework covering a large set of resources. A virtualization management approach is then described, with a proposed ontology for ICT resource focusing on energy consumption. A resource discovery mechanism is next provided where queries are built in a flexible and extensible manner with the help of a Bayesian network model and a knowledge base. We also provide an example of using the proposed resource description and discovery framework in the GreenStar Network in order to reduce GHG emissions. Finally, we conclude the chapter and present future work.

6.2 Low-Carbon Network and Resource Management

The research presented in this chapter is positioned in a Canada-international project, aimed at building the first nationwide network, the GreenStar Network, which is powered only by green energy sources. The project is inspired from the carbon-neutral approach proposed in [3] and from the emerging need of a green energy distribution wide area network model for establishing a standard carbon protocol for the ICT industry. The key objective of the GreenStar Network project [2] is to create a pilot and a testbed environment from which to derive best practices and guidelines to follow when building low-carbon networks. The idea behind the Green-Star Network project is that a neutral carbon network must consist of data centers built in proximity to clean power sources, and user applications will be moved to be executed in such data centers (Figure 6.1). Such a network must provide an ability to migrate entire virtual machines (routers and servers) to alternate data center locations. An underlying communication network is supported by a high-speed optical layer having up to 1,000 Gbps bandwidth capacity. Note that optical networks have a modest increase in power consumption, especially with new 100 G and 1,000 G waves, in comparison to electronic equipment such as routers and aggregators [3]. The key technology of the GreenStar Network is virtualization. The migration of a data center over network nodes is indeed a result of a combination of server and network virtualizations as virtual infrastructure management.

Figure 6.1 shows the connection plan of the GreenStar Network. The Canadian section of the network has the largest deployment of six nodes powered by sun, wind, and hydroelectricity; each node represents a data center. The core GreenStar Network is connected to the European green nodes in Ireland (HEAnet), Spain (i2CAT), and the USA (CalIT2, through

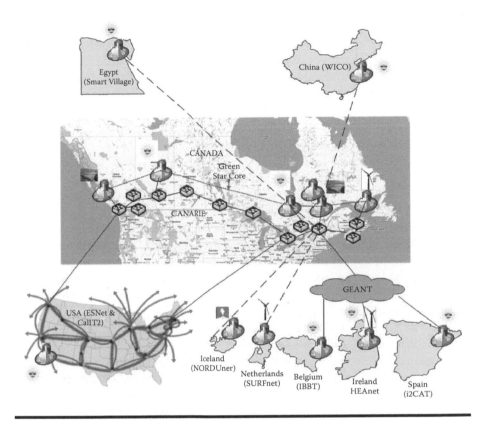

Figure 6.1 The GreenStar Network.

ESNet). Other nodes in Belgium (IBBT), the Netherlands (SURFnet), Iceland (NORDUnet), China (WiCo), and Egypt (Smart Village) are also planned to connect to the network.

In such a network, data centers are smart grids that are capable of turning on/off and migrating application data to remote locations in order to utilize available green energy sources. This requirement is achieved by a flexible and robust resource description and discovery framework where energy control a is key characteristic. Such a framework is not yet available in current grid resource management frameworks. Indeed, most of current grid resource description and discovery research focuses on XML structured data, such as the Globus Toolkit's Monitoring and Discovery System (MDS4) [4,6,7], in order to enable Web service interfaces. Nevertheless, a semantic relationship among resources is not fully taken into account, particularly as it is related to energy consumption interpretation, which results in big challenges for building low-carbon networks. Additionally, the current method based on collecting and publishing aggregated information can decrease search performance. Some research has suggested handling the semantic aspect based on the Web Services Resource Framework (WSRF) specification [9]. However, WSRF will unlikely be appropriate for managing heterogeneous environments. In addition, current methods usually maintain a large number of sophisticated components, which impose several challenges for implementation and maintenance.

Until very recently, many grid resource management systems still did not support semantic resource description and discovery. For example, Gridbus broker [10], Gridway [11], and the Monitoring and Discovery System (MDS) [12] simply use Globus middleware services to gather grid resource information; thus, it is impossible to understand the semantic relationship between the available resource information and the requested information. Their resource discovery mechanism is based on conventional keyword matchmaking. Some other research has proposed various support levels for semantic relationships among resources [13], for example, using ontology to describe and select web-based services used in specific domains, like life sciences and aeronautical engineering [14]. However, no virtualized management approach is proposed.

One of the key challenges for a resource description and discovery system is automation. Such a system must be able to integrate knowledge from resource providers, service providers, and users in order to enrich system databases and to empower searching performance. In [9], authors proposed a completed framework for ICT resource management in grid-based networks using a predefined ontology. Their research, however, focuses on resource utilization and searching performance. Energy-related aspects and semantics expressing relationships among resources in terms of energy consumption interdependence are not taken into account.

6.3 Virtualized Management

The GreenStar Network is built with multiple layers, resulting in a large number of resources to be managed. In order to scale management activities along with the growth of grid resources, virtualized management has been proposed for service delivery regardless of the physical location of the infrastructure, which is determined by resource providers. This keeps complex underlying services hidden inside the infrastructure provider. Resources are allocated according to user's need, and hence highest utilization and optimization levels can be achieved. During the service, the user owns and controls resources as if he were the owner. Therefore, users will be able to run their application in a virtual infrastructure powered by green energy sources.

Such a service provisioning model has three layers: infrastructure provider, service provider, and end users (Figure 6.2). End users send jobs

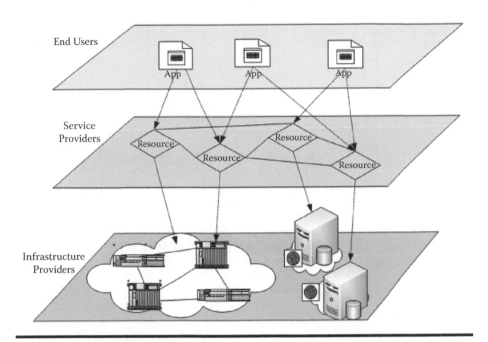

Figure 6.2   Virtualized service provisioning.

to infrastructure providers through service providers and get back results. The service provider layer consists of several functionalities, such as resource registry, reservation, work scheduling, resource aggregation, data routing, and so on.

The virtualized management approach for grid resources requires a flexible and extensible resource description and discovery support. The interdependence between a service provider and an infrastructure provider, as well as resource sharing and aggregation capabilities, must be represented semantically. One of the appropriate solutions for such a description is using a particular ontology.

6.4 ICT Energy Consumption Ontology

Ontologies (often also referred to as domain model) are generally defined as a *"representation of a shared conceptualization of particular domain."* As a traditional textual keyword-based approach is not efficient in resource description and discovery regarding energy consumption characteristics, an ontology should be provided to represent relationships of resources in this domain.

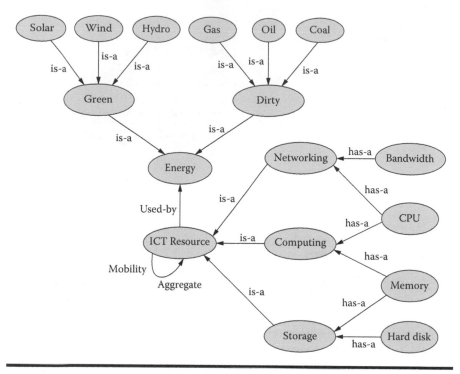

Figure 6.3   ICT energy consumption ontology.

The proposed ontology includes knowledge models for ICT elements, hosting application, and energy. These models are mapped into classes and properties assigned to these classes. The ontology contains a hierarchy of semantically linked elements and describes their attributes and capabilities. As shown in Figure 6.3, the proposed ontology is a combination of two large domains: ICT and energy. It includes the following classes:

The ICT resources class is a generic type of ICT resources. It has some common characteristics, like name, ID, location, and power consumption. ICT resources has three subclasses: (1) computing class represents all computing resources, such as servers, PCs, or handheld devices; (2) networking class represents all resources used for interconnecting ICT devices, such as routers, switches, hubs, optical cross-connects, and multiplexer-demultiplexer, and (3) storage class represents all storage resources, such as hard disk, CDs, and multirack storage devices.

There is a relationship between ICT resources and energy that defines the kind of energy consumed by ICT devices, e.g., hydroelectricity, solar, wind, or fuel. A relationship between ICT resource and energy also defines the amount of energy consumed by an ICT device during a time unit (i.e., an hour). ICT resources is associated with a concept, named mobility, to define whether the resource can be portable or not. A resource may also have an aggregation capability, which defines whether the resource can be aggregated in order to achieve a given task.

An energy source can be green energy or dirty energy. Green energy can be solar, wind, or hydroelectricity. Dirty energy includes natural gas, heating oil, or coal. Each energy class has an associated cost. Dirty energy is more costly than green energy. In the context of the GreenStar Network project, we define energy cost as CO_2 emissions of an energy unit.

These classes are implemented using RDF [8], which is considered the best representation for ontology. Unlike traditional databases where tables are fixed and cannot handle semantic aspects, RDF supports the semantic representation and information is mapped directly to models, making the method flexible and extensible. Reasoning is also enabled in RDF, using a triple object-attribute value: an object O has an attribute A with value V [8].

6.5 Proposed System Architecture

Figure 6.4 shows the global architecture of the proposed resource description and discovery framework used for the GreenStar Network. There are two stacks in the figure. The resource provider stack aims at determining ontology concepts and description for exposed resources. On the other hand, the end user stack is dedicated for searching appropriate resources according to each user request.

The system maintains a local index to register all resources. Resource providers define description and capabilities for their resources. One or several keywords can be used for each resource to facilitate indexing and searching.

The semantic analyzer performs information processing based on resource descriptions in order to determine the type of resources and associates resources to appropriate ontology concepts. The semantic analyzer may also extract keywords from a description. Newly found keywords or concepts will be added to the semantic analyzer knowledge base. Finally, resource registry service is invoked to register resources in the RDF data model base.

The end user stack begins with a request for resources. The query is analyzed and keywords are extracted and processed by the semantic analyzer. Based on its knowledge base, the semantic analyzer determines the required resources and their locations. If a resource is available, it will be triggered. The proposed system focuses on the energy aspects. Therefore, the set of resources will be returned, taking into account the minimal GHG emissions possible.

In order to provide a powerful search method based on the proposed ontology, we use a Bayesian semantic graph, which combines a semantic

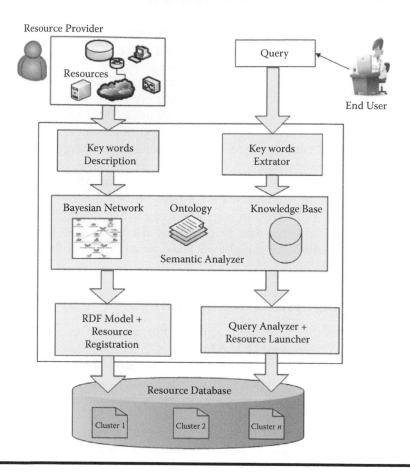

Figure 6.4   Global architecture.

inference and a probabilistic one. The key idea is to integrate semantic links and reasoning rules, by processing keywords and taking into account the context, which determine the cluster the resource belongs to. As proposed in our ontology, a resource will be placed in one of the three categories: computing, storage, and network. However, a user query like "RAM = 64 Mb, IP address = 10.0.0.1," may lead to a confusion, because the result can be a server or a router. If further information, such as "bandwidth = 1G," is added, a more accurate network resource could be found. The Bayesian semantic analyzer is used to deal with such confusion. It processes all the words in each resource description or user query and calculates the probability that resource belongs to each cluster. Such a mechanism improves significantly the search operation.

Generally, providers and end users use different manners to describe resources and express their requests. The knowledge base is used to find

Table 6.1   Example of Clusters

CPU	RAM	Bandwidth	Word Concept
0.99	0.97	0.47	Compute
0.47	0.50	0.20	Storage
0.60	0.60	0.99	Network

the similarity between these different keywords. It is built based on a thesaurus dictionary. When the input keyword is not found, a synonym-based searching will be triggered. Thus the searching operation is more efficient, since many of the bad formed requests or poor descriptions can be addressed.

In order to define the concepts and clusters for a Bayesian network, a probabilistic table is built, as proposed in [5], which assigns joint probability values. The assignments are based on expert judgment and may be improved over time. The probability table is a matrix, as shown in Table 6.1, which has as rows words and columns concepts. A value in each matrix cell represents the probability of the word to be involved in the concept.

When a resource has been analyzed and its keywords have been determined, it is represented by an RDF graph. This operation is achieved using an RDF request language, such as SPARQL [18] in our implementation.

In the user stack, user queries are also analyzed in order to extract keywords. Similarly, probability is calculated for each keyword, and then target concept and cluster are determined. A substituted synonym can be used if no cluster is found. Finally, an appropriate resource will be obtained from the resource database. In the worst case, where no resource is found, users will be suggested to use alternate keywords.

6.6 Carbon-Aware Resource Discovery

The proposed resource discovery mechanism aims at minimizing the CO_2 emissions produced by ICT resources for each service request. For example, a user request for 25 CPU, 2 GHz speed each, can be achieved by many available servers distributed across the GreenStar Network. However, at a given period of time, only some of them are powered by green energy sources (i.e., solar energy is not available during the night). Thus, the proposed resource discovery mechanism tries to find a list of optimal resources in terms of CO_2 emissions. The problem can be formulated as follows.

Given N resources in the network, each resource R_i has a set of capabilities $\{C_{ij}\}$ and a power P_f. Resource R_i is powered by an energy source E_m, which is associated with a CO_2 emission factor E_{mi}.

A user request for a set of L capabilities: $Q = \{Q_1\}$, where $Q_1 = \sum C_1$ is a sum of capabilities C_l.

The resource discovery engine has to decide which resource in the network should be assigned for the request Q. The following matrix of binary variables represents the resource selection result:

$$X_i = \begin{cases} 1 & \text{if } R_i \text{ is selected} \\ 0 & \text{otherwise} \end{cases} \tag{6.1}$$

The mathematical optimization problem is to optimize:

$$\min \sum_i^N x_i \times P_i \times E_{mi} \tag{6.2}$$

subject to:

$$\sum_{i,l \psi L}^N x_i \times C_{il} \geq Q_l \tag{6.3}$$

This formulation is referred to as the original mathematical optimization problem. Since the objective function is nonlinear and there are nonlinear constraints, the optimization model is a nonlinear programming problem with binary variables. Generally, a solution for this problem cannot be obtained by mathematical programming solvers. In this chapter, we use a simple greedy algorithm to determine solutions for each user request, assuming a relatively small number of resources in the network (i.e., order of 10,000). The algorithm first tries to find a list of resources that can meet user requirements. The resources in the list will then be sorted according to energy sources. They will be picked from the top of the list until the required number is filled.

6.7 Experimental Results

In the current GreenStar Network, there are 6 different data centers powered by different energy sources. According to the Canadian GHG Registries [17], electricity emission factor (i.e., tons of CO_2 per Kwh) varies from province to province. For example, Québec has the lowest emission factor and Alberta has the highest one, due to the fact that AB uses fossil power while QC uses hydroelectricity.

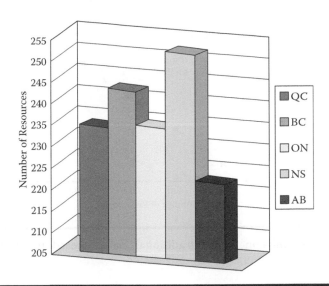

Figure 6.5   Distribution of resources over network nodes according to a user request (CPU = 2 GHz, RAM $\geq$ 1,024 MB, OS = Ubuntu).

In our experiments, a user query such as "CPU = 2 GHz, RAM $\geq$ 1, 024 MB, OS = Ubuntu" may return a bunch of servers. A user application service, like GeoChronos [16], developed by the University of Calgary, requires 48 servers with that configuration. Figure 6.5 shows the distribution of resources found for such a user request. In a traditional resource discovery, when the GHG emission aspect is not taken into account, the first 48 servers in the finding results will be returned regardless of their locations or energy sources. It would lead to a nonoptimal resource allocation scheme from the point of view of ecologists. The proposed resource discovery engine will try to allocate resources powered by greenest energy sources; thus the GHG emission is minimal.

In reality, resources located in Québec are the most efficient in terms of GHG emissions because they are all powered by green energy (i.e., hydroelectricity). Unfortunately, their number and availability make it impossible to always allocate resources in Québec for all user requests. Therefore, a best-effort discovery must be provided in order to find an optimized possible resource list.

Figure 6.6 compares GHG emissions resulting from the traditional and proposed resource discoveries. The amount of CO_2 emissions is calculated based on [17]. As shown in the figure, when the number of required resources is 900, we may save up to 128 credits of CO_2. The carbon credit savings is significant, particularly when the number of resources allocated increases, which is very useful for large IT companies when a carbon tax is imposed (i.e., which is currently the case in British Columbia).

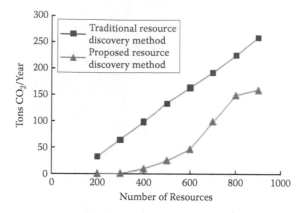

Figure 6.6 CO_2 **emission in the traditional and proposed resource discovery schemes.**

6.8 Conclusion

In this chapter, we have presented a framework for resource description and discovery in large-scale smart grid networks, leveraged by the aim of reducing GHG emissions of ICT resources. Results of this research will be used for GreenStar Network, the first nationwide network powered entirely by green energy. An energy-oriented ontology has been proposed to support semantic descriptions of ICT resources in terms of energy consumption. Taking into account the inefficiency of current resource description and discovery frameworks, a web semantic approach has been deployed for resource description and persistence. The resource discovery engine we developed integrates a knowledge base and a Bayesian network in order to improve searching performance. GHG reductions resulting from experimental evaluation suggest potential applications of the research in large-scale grid networks.

Our future work includes resource control and migration, driven by the change of the green energy sources, such as solar and wind. The resource discovery algorithm will also be improved, regarding the increasing number of resources in the network. Optimization mechanisms will be investigated, with respect to combined objective functions, such as QoS and energy together.

Acknowledgments

We thank CANARIE for funding the GreenStar Network project under its G-IT pilot program. The authors acknowledge all GreenStar Network partners for their contribution in this project, from network implementation to carbon assessment and quantification.

References

1. H.D. Saunders, The Khazzoom-Brookes Postulate and Neoclassical Growth, *Energy J.*, 13(4), 130–148, 1992.
2. The GreenStar Network Project, *http://greenstarnetwork.com*.
3. S. Figuerola, M. Lemay, V. Reijs, M. Savoie, B. St. Arnaud, Converged Optical Network Infrastructures in Support of Future Internet and Grid Services Using IaaS to Reduce GHG Emissions, *J. Lightwave Technol.*, 27(12), 1941–1946, 2009.
4. S.M. Pahlevi, I. Kojima, Towards Automatic Service Discovery and Monitoring in WS-Resource Framework, in *Proceedings of Semantics, Knowledge and Grid SKG '05*, 2005, p. 106.
5. K. Kyoung-Min, H. Jin-Hyuk Hong, C. Sung-Bae, Intelligent Web Interface Using Flexible Conversational Agent with Semantic Bayesian Networks, in Proceedings of Next Generation Web Services Practices NWeSP'05, 2005, pp. 22–26.
6. A. Chervenak, J.M. Schopf, L. Pearlman, S. Mei-Hui, S. Bharathi, L. Cinquini, M. D'Arcy, N. Miller, D. Bernholdt, Monitoring the Earth System Grid with MDS4, in *Proceedings of e-Science and Grid Computing e-Science '06*, 2006, p. 69.
7. J.M. Schopf, I. Raicu, L. Pearlman, N. Miller, C. Kesselman, I. Foster, M. D'Arcy, Monitoring and Discovery in a Web Services Framework: Functionality and Performance of Globus Toolkit MDS4, *J. Physics*, 46, 521–525, 2006.
8. S. Decker, S. Melnik, V. Van-Harmelen, D. Fensel, M. Klein, J. Broekstra, M. Erdmann, I. Horrocks, The Semantic Web: The Roles of XML and RDF, *IEEE Internet Comput.*, 4(5), 63–73, 2000.
9. B.R. Amarnath, T.S. Somasundaram, M. Ellappan, R. Buyya, Ontology-Based Grid Resource Management, *Software Practice and Experience*, 39(17), 1419–1438, 2009.
10. S. Venugopal, R. Buyya, L. Winton, A Grid Service Broker for Scheduling E-Science Applications on Global Data Grids, *Concurrency and Computation: Practice and Experience*, 18(6), 685–699, 2006.
11. E. Huedo, R.S. Montero, I.M. Llorente, The Gridway Framework for Adaptive Scheduling and Execution on Grids, *Scientific International Journal for Parallel and Distributed Computing*, 6(3), 1–8, 2005.
12. K. Czajkowski, S. Fitzgerald, I. Foster, C. Kesselman, Grid Information Services for Distributed Resource Sharing, in *Proceedings of the Tenth IEEE International Symposium on High-Performance Distributed Computing* (HPDC-10), 2009.
13. O. Corcho, P. Alper, L. Kotsiopoulos, P. Missier, S. Bechhofer, C. Goble, An Overview of S-OGSA: A Reference Semantic Grid Architecture, *Journal of Web Semantics*, 4(2), 102–115, 2006.
14. M.J. Murphy, M. Dick, T. Fischer, Towards the Semantic Grid: A State of the Art Survey of Semantic Web Services and Their Applicability to Collaborative Design, Engineering, and Procurement, *Communications of the IIMA*, 8(3), 11–24, 2008.

15. W. Binder, N. Suri, Green Computing: Energy Consumption Optimized Service Hosting, *in SOFSEM 2009: Theory and Practice of Computer Science*, vol. 54, 2009.

16. C. Kiddle, GeoChronos: A Platform for Earth Observation Scientists, *OpenGridForum* 28, 2010.

17. LivClean Carbon Offset Solution, *How Is This Calculated?*, http://www.livclean.ca.

18. E. Prud'hommeaux, A. Seaborne, *SPARQL Query Language for RDF*, W3C Recommendation, 2008. http://www.w3.org/TR/rdf-sparql-query/.

COMMUNICATIONS AND NETWORKS IN SMART GRIDS

Chapter 7

An Optimum Method to Design Distributed Electric Power Supply and Communication Networks

Susumu Yoneda

Contents

Existing grid infrastructures are optimized for centralized carbon-based energy sources and long-distance power transmission. Concerns about climate change and greenhouse gas (GHG) emissions would lead us to increase our dependency on distributed small-sized low-carbon energy sources coupled with short-distance power transmission. Owing to the low energy densities as well as the distributed layout of low-carbon sources, the existing grid infrastructure might not be suitable or efficient. Therefore, issues pertaining to the design of a distributed low-carbon infrastructure, i.e., a microgrid, are investigated, and an optimal design method is proposed.

A distributed electric power supply system can function independently. With respect to its fluctuating power generation and smooth migration from the existing grid architecture, we propose a centralized conventional architecture laid over many new small distributed electric power supply systems with communication systems, which can control multiple grid systems by utilizing various data acquired from the distributed systems. Furthermore, the proposed optimization model allows supply of a constant base power in a centralized manner; therefore, we can expect to achieve efficient and centralized power generation.

7.1 Introduction

In light of the problems such as climate change and greenhouse gas (GHG) emission, the smart grid can be viewed as one of the solutions to promote the use of low-carbon energy sources such as solar and wind. The smart grid utilizes information communication technologies (ICT) so that energy generation, consumption, and transshipment can be monitored and controlled.

The smart grid contains many small distributed power supply systems, i.e., microgrids, and the design of these microgrids is a key factor to be considered while designing the smart grid. For designing a microgrid, a mathematical optimization model is developed in this chapter, and solutions to some of the key issues pertaining to the design of a microgrid are presented for a simple sample model. Key issues include electric power transfers among neighbors, and the impact of these transfers on the required power storage capacity and the low-carbon power generation capacity. These are obviously important elements in microgrid design. Furthermore, the proposed optimization model allows supply of a constant base power in a centralized manner; therefore, we can expect to achieve efficient and centralized power generation. Fluctuating power demands are absorbed by low-carbon energy sources with power storages.

7.2 ICT and Climate Change

The standardization of ICT in light of climate change is an ongoing activity at the International Telecommunication Union (ITU-T) [1–4]. There are two aspects with regard to the mitigation of GHG emission and ICT use. ICT itself consumes electricity, and emits some GHG. As such, by introducing new designs and technologies for ICT, GHG emissions generated by ICT can be reduced. ICT also provides alternatives to conventional methods of conducting some business and personal activities. For example, the use of a TV conference may substitute a business trip. Because ICT is responsible for only 2% of the total global GHG emissions, the use of ICT may significantly reduce GHG emissions.

In any case, it is important to establish a globally agreeable methodology to determine the impact of ICT use. To this end, the ITU-T is working on standardization so that the impact of ICT use can be accurately and fairly evaluated.

The methodology adopted for standardization may consider the impact of the electricity generated by low-carbon technologies. The basic approach focuses on saving electricity and reducing energy usage so that GHG emissions can be reduced. However, in this chapter, a proactive approach is adopted; i.e., the use of solar and wind energy is increased. To promote this proactive approach, microgrid design is investigated.

7.3 Microgrids

Solar and wind power are low-energy-density sources. Therefore, a microgrid cannot cover a wide area by itself. In order to cover a large area, many interconnected microgrids in conjunction with an existing core power/communication network infrastructure are required. Figure 7.1 shows a conceptual picture of the smart grid consisting of the microgrids and core infrastructure [5].

Note that two types of interfaces are shown in Figure 7.1 as black and white small circles. These interfaces should be standardized so that connectivity is easily realized, and technological developments within a particular microgrid do not affect its neighboring microgrids or the core network and vice versa.

The microgrids consist of several components, and the design criteria associated with these components can be addressed in the following:

■ Solar and wind power generation capacities
■ Power storage capacity
■ Fixed or moveable, i.e., electric car

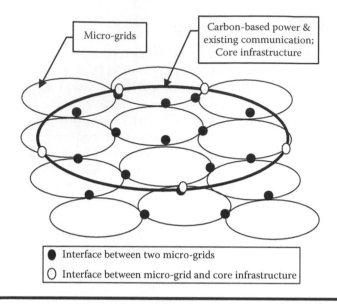

Figure 7.1 Microgrid and core infrastructure.

- Number of neighbors
- Size of the microgrid
- Frequency of control actions
- Frequency of data monitoring and collection
- Amount of carbon-based power supply

The microgrid warrants a new design, and to physically construct it, numerical values of the above-mentioned components are calculated. In this chapter, an optimization model is developed, and optimum values of various components of a simple sample problem are computed and analyzed.

7.4 Motivation behind Microgrids

The amount of power generated by solar and wind power generation technologies depends on natural phenomena such as sunlight and wind. This implies that we cannot effectively control the power generated by these sources. Therefore, a power storage facility is required at or near the power generation site. Ideally, every solar or wind power generator must be near an electric power storage facility. In reality, however, considering the construction cost, this might be difficult.

Let us assume that a solar and wind power generation site has the required storage facilities, and its adjacent solar and wind power generation sites also have similar facilities. The amount of electric power stored and power consumption at each site will be different; even if a site has almost no power in its storage and the storage of the adjacent site is full, the site requires some electric power from the public power supply, because there is no link between the two adjacent sites. Therefore, a microgrid, which provides this link, is needed so that the surplus stored power can be shared with the neighbors. This microgrid transmits not only electric power but also control information, so that the level of storage and consumption at the adjacent site is available, and therefore a suitable amount of electric power can be supplied to the neighbors.

Figure 7.2 shows infrastructure independent of carbon-based power supply. Sites A and B are powered by carbon-based, solar, and wind power supply systems. As shown in Figure 7.2, the electric power level of the storage at site A is low. On the other hand, the power level of the storage at site B is high. In the absence of a microgrid, site A requires carbon-based

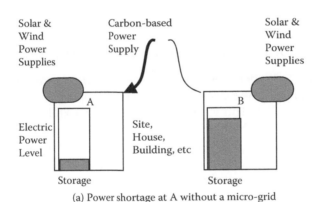

(a) Power shortage at A without a micro-grid

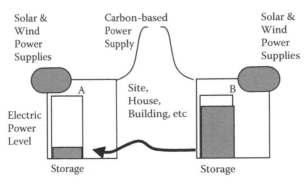

(b) Power shortage at A with a micro-grid

Figure 7.2   Infrastructure independent of carbon-based power supply.

power supply. However, with a microgrid, site A can receive power from site B. Consequently, site A can function independently of the carbon-based power supply.

Because of solar and wind power generation, a small cluster of connected sites is considered, such that power generated by and stored at a site could be shared with neighboring sites. This cluster becomes a microgrid.

Hydroelectricity and nuclear power are also low-carbon energy sources. However, at this point, we consider these sources to be a part of the core existing infrastructure. A separation of hydro power and nuclear power from carbon-based powers requires further study.

7.5 Mathematical Model

The design components of the microgrid are analyzed and optimized for a simple sample problem. Here, a multiobjective programming model is used [6].

7.5.1 Descriptions of Variables Used in the Model

Figure 7.3 shows the decision variables and the known variables used in the model. The values of the decision variables are optimized after solving the multiobjective model. The values of the known variables are given before solving the model, and they are input data for the model.

Note that Cti can be a decision variable when its given upper and lower bounds are considered. When introducing cost or price coefficients associated with carbon-based and solar/wind power supplies, the results could

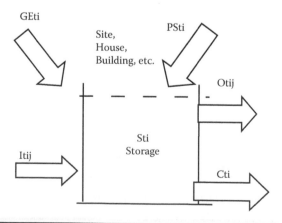

Figure 7.3   Variables used in the mathematical model.

address economic aspects of using a combination of carbon-based and so-lar/wind energy sources. The optimization of the economic aspect requires further study.

7.5.2 Model Formulation

Here, we have used a multiobjective programming model. Usually, the objective of a model is to minimize the cost. In this chapter, we aim to minimize the maximum requirements of critical capacities, i.e., solar/wind power generation, storage, and carbon-based power supply. These criti-cal aspects will certainly have an impact on the construction cost of the microgrid.

There are three objectives. The first objective is to minimize the max-imum requirement of stored power. The second objective is to minimize the maximum requirement of solar/wind power generation. The third ob-jective is to minimize the maximum requirement of carbon-based power supply. These three objectives are, respectively, mathematically described as follows:

- ■ Minimize S
- ■ Minimize GE
- ■ Minimize PS

Note that S, GE, and PS represent the maximum values of the variables of stored power, solar/wind power generation, and carbon-based power generation, respectively.

Now, the constraints on the model are as follows.

The first constraint set indicates an electric power balance at site i. Power inputs are solar/wind power generation, carbon-based power sup-ply, and power transfer from the adjacent sites. Power outputs are the difference between the power storages at the end and the beginning of period t and power transfers to the adjacent sites. The difference between the inputs and outputs should be greater than or equal to the amount of power consumptions at site i.

$$S_{t-1i} - S_{ti} + PS_{ti} + GE_{ti} + \sum_{j \varepsilon Ji} I_{tij} - \sum_{j \varepsilon Ji} O_{tij} \geq C_{ti} \quad \forall t, i$$

The second constraint set indicates that the power stored at site j should be greater than or equal to the power transferred from site j to site i. $j \varepsilon Ji$ indicates that site j is connected to site i. As such, site i can receive and transfer electric power from and to site $j \varepsilon Ji$.

$$S_{t-1j} - I_{tij} \geq 0 \qquad \forall t, i, j \varepsilon Ji$$

The third, fourth, and fifth constraint sets obtain the maximum values of the variables of storage, carbon-based power supplies, and solar/wind power generation, i.e., S, PS, and GE, respectively.

$$S - S_{ti} \geqq 0 \qquad \forall t, i$$

$$PS - PS_{ti} \geqq 0 \qquad \forall t, i$$

$$GE - GE_{ti} \geqq 0 \qquad \forall t, i.$$

Finally, all decision variables are nonnegative.

The model size is dependent on t and i. When t and i are large, the interval between two consecutive control actions decreases, and a larger portion of the microgrid can be analyzed. Therefore, all design components addressed in Section 7.3 can be resolved using this optimization model.

7.5.3 Simple Sample Problem

Next, a simple sample problem is solved. The model has only four sites, each of which is connected to two other sites. All sites can receive electric power from a carbon-based power supply, and they also have their own storages. Since all four sites are connected, $Otij$ (power transfer from site i to site j during period t) can be substituted with $Itij$. For example, the power transfer $Ot12$ from site 1 during the period t is exactly the same as the power transfer $It21$ from the viewpoint of site 2. Figure 7.4 shows the simple sample problem. In this problem, we consider 10 periods; each of these may represent an hour, a day, etc. We also assume that the total power consumption at each site is represented by 100, and that no unit is associated with this power consumption. This implies that the sum of power consumption in the 10 periods is 100% at each site.

The objectives of this problem are the same as the three objectives addressed during model formulation, i.e., minimizations of S, GE, and PS.

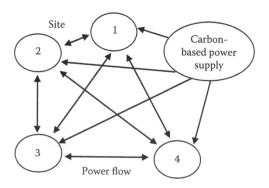

Figure 7.4 Simple sample problem.

Table 7.1 Power Consumption Patterns

Period	Identical Demand Pattern Site 1/2/3/4	Different Demand Pattern Site 1	Site 2	Site 3	Site 4
1	5	5	5	20	20
2	5	5	5	10	10
3	10	10	10	10	10
4	10	10	10	5	5
5	20	20	20	5	5
6	20	20	20	5	5
7	10	10	10	5	5
8	10	10	10	10	10
9	5	5	5	10	10
10	5	5	5	20	20
Total	100	100	100	100	100

Out of the various methods available for solving a multiobjective problem, we used a weighting method. A computational result shows the trade-offs among these three objectives. Because this is a three-dimensional problem, the trade-offs will be represented by a contour surface on a three-dimensional graph.

Power consumption at each site during period t is a given variable. The other variables are decision variables. Two demand patterns of power consumption are assumed in this chapter: identical and different. An identical pattern is where all four sites show an identical power consumption pattern given in Table 7.1. The patterns in the case of different demand are also listed there. In both these patterns, the total power consumption at each site is 100.

7.5.4 Computational Results

By changing the weights of each of the three objectives, we can obtain several trade-offs among them. For example, the three objectives can be described as a minimization of A × S + B × GE + C × PS. When A ≥ B or C, the value of S will decrease or approach zero. A similar approach can be adopted for GE or PS. Thus, we can obtain the trade-offs among the three objectives.

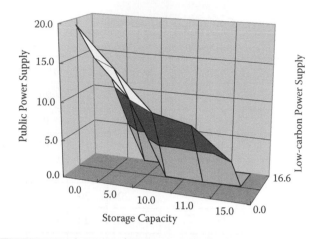

Figure 7.5 Contour surface of the feasible region for identical demand pattern.

Figure 7.5 shows the trade-off contour surface for the identical demand case. The feasible solutions are located in the upper region of the contour surface. Lower regions of the contour surface indicate the infeasible solution space. For example, the point represented by S = 0, GE = 0, and PS = 0 is infeasible. This is intuitively correct because here, not every power consumption demand can be satisfied.

The three objectives are the maximum required capacity and the two types of power supply at a site, i.e., storage capacity, low-carbon power supply, and carbon-based power supply. Here, we also refer to the carbon-based power supply as the public power supply. Table 7.2 lists the results of the objective values in both the identical and different demand patterns. For all results, the sum of the three objective values is 20. In the different demand pattern, the storage capacity value for all solutions given in Table 7.2 is 5. On the other hand, the storage capacity value for the identical demand pattern is zero. This means that the identical demand pattern requires a higher public power supply than the different demand

Table 7.2 Results of Identical and Different Demand Pattern Problems

Identical Demand Pattern			Different Demand Pattern			
Storage	Low-Carbon	Public Supply	Storage	Low-Carbon	Public Supply	Total
0	0	20	5	0	15	20
0	5	15	5	6	9	20
0	10	10	5	10	5	20
0	15	5	5	15	0	20

pattern does. With respect to GHG emission, a lower value of the public power supply is preferable. Consequently, a microgrid should include at least some sites that have the different demand pattern.

7.6 Analyses of Results

Since the peak value of power consumption at a site during one period is 20, we can assume that the minimum value of the sum of the three objectives will be 20, which is verified by solving the optimization problem. Even without solving the optimization problem, we conventionally know that the required power capacity depends on the peak power consumption. However, in a microgrid situation, we have three completely different elements for the power capacity and generations: power storage, low-carbon power generation, and public conventional power supply. The optimal mixture of these three elements is the key factor for designing a microgrid. The mathematical model proposed in this chapter will provide a foundation for solving this designing problem.

As mentioned earlier, two demand patterns are considered in this simple sample problem: identical and different. The sum of power consumption at each site for 10 time periods is identical for both patterns. However, as shown in the previous section, the optimum solution sets of the different demand pattern have a lower value of the public power supply than those of the identical demand pattern. This is simply because of the power transfers between two sites. If one site has surplus power in its storage, a part of it can be transferred to another site that is facing a power shortage. Consequently, the public power supply can be reduced. Then, for the identical demand pattern, all sites have an identical pattern of demands. For each site, the peak of power consumption occurs in the same period, and power transfers may not occur in such a situation. Table 7.3 lists the number of links having nonzero power transshipment.

Table 7.3 Number of Links with Nonzero Power Transshipment in the Solutions

Identical Demand Pattern				Different Demand Pattern			
Storage	Low-Carbon	Public Supply	Nonzero Links	Storage	Low-Carbon	Public Supply	Nonzero Links
0	0	20	0	5	0	15	5
0	5	15	0	5	6	9	21
0	10	10	0	5	10	5	5
0	15	5	0	5	15	0	5

Table 7.4   Solution for Constant Public Power Supply

Different Demand Pattern			
Storage	Low-Carbon	Public Power Supply	Total
2.5	7.5	10	20
5	7	8	20

At the least, power transfer capabilities among adjacent sites will lead to a reduction in the carbon-based public power supply. This indicates that the connections among the various sites in the microgrid are effective and important design criteria.

In reality, each household or commercial building has a unique power consumption pattern. If a long time interval is chosen, its power consumption may appear similar, and the model will consider this problem as the identical demand pattern. However, the different demand pattern is preferable in such a case. Therefore, a smaller interval should be applied for collecting data and controlling devices within a microgrid.

In this simple sample problem, GE is a decision variable and it does not have an upper limit. However, in reality, depending on the geographical location, GE will be severely restricted. When a target location is chosen, the upper limit of GE can be added to the model. PS is also a decision variable; it too has an upper limit. Furthermore, from a power company's point of view, a constant value of power supply is preferable. Then, the power company's operation would be simplified. This type of constraint is added to the model. Table 7.4 shows the results.

In Tables 7.2 and 7.3, the values of storage, low-carbon power generation, and public power supply are the minimized maximum values of those objective functions throughout 10 time periods. In Table 7.4, the values of the storage and the low-carbon power generation are also the minimized maximum values. However, the values of the public power supply listed in Table 7.4 are constant values throughout the 10 time periods. This means that the same amount of public power supply is required in each time period. Therefore, the proposed model can permit the conventional power company to operate a constant power supply in every time period.

7.7 Conclusion

The smart grid has two important benefits: effective use of solar and wind energy sources and efficient collection of data from and control of various sensor devices in the smart grid. For designing the smart grid, this chapter proposes a conventional grid architecture laid over many new, small distributed microgrids, which facilitate power transfers among adjacent sites. By solving an optimization problem, we show the effectiveness of the power transfers among the adjacent sites.

The microgrid also consists of power supply and monitoring/control parts. The former is further divided into two parts: solar/wind power supplies and carbon-based power supplies. To promote the solar/wind power supplies, we propose a mathematical optimization model that can yield the values of some critical design parameters, such as the required storage capacity and the solar/wind and carbon-based power supply capacities. The proposed optimization model allows supply of a constant base power in a centralized manner; therefore, we can expect to achieve efficient and centralized power generation. Fluctuating power demands are absorbed by low-carbon energy sources with power storages. Using this model, we show that a constant carbon-based power supply can be achieved in every time period.

With regard to the required control frequency in the microgrid, when there is a difference in power consumption among sites, we can select a suitable time interval for the control actions. The proposed model can also change the number of sites within a subnetwork of the microgrid. The number of sites will be an important factor in designing the monitoring and control aspects of the microgrid. The simple model presented in this chapter is flexible enough to incorporate various future requirements in terms of power supply and consumption.

Acknowledgments

The author appreciates Wolfram Research for providing a free demo license of Mathematica 7. The optimization problems were solved using Mathematica 7.

References

1. ITU ICT and Climate Change Focus Group Report, Deliverable 1, Definitions, ITU Telecommunication Standardization Advisory Group Meeting, Geneva, April 28–30, 2009.
2. ITU ICT and Climate Change Focus Group Report, Deliverable 2, Gap Analysis and Standards Roadmap, ITU Telecommunication Standardization Advisory Group Meeting, Geneva, April 28–30, 2009.
3. ITU ICT and Climate Change Focus Group Report, Deliverable 3, Methodologies, ITU Telecommunication Standardization Advisory Group Meeting, Geneva, April 28–30, 2009.

4. ITU ICT and Climate Change Focus Group Report, Deliverable 4, Direct and Indirect Impact of ITU Standards, ITU Telecommunication Standardization Advisory Group Meeting, Geneva, April 28–30, 2009.
5. Kirthiga, M.V., and Daniel, S.A., Optimal Sizing of Hybrid Generators for Autonomous Operation of a Microgrid, 2010 IEEE 26th Convention of Electrical and Electronic Engineering in Israel, 2010, pp. 864–868.
6. Cohon, J.L., *Multiobjective Programming and Planning*, Academic Press, New York, 1978.

Chapter 8

A Smart Grid Testbed: Design and Validation

Wen-Zhan Song, Debraj De, and Gang Lu

Contents

The concept of smart grid is a form of electricity network that will deliver electricity from distributed suppliers to consumers, with the help of two-way digital communications that will control the energy storage and distribution process. This will save energy, reduce costs, and increase reliability in the presence of failures. The transformation of an existing traditional centralized electricity grid into a state-of-the-art smart grid will need innovation in a number of dimensions, such as seamless integration of renewable energy sources, management of intermittent power supplies, real-time demand response, dynamic energy pricing strategy, self-healing for disruption resilience, information communication infrastructure, etc. The grid configuration will need to change from the centralized power broadcasting network into a more distributed, dynamic, and robust network with two-way energy transmission. Another necessary component for smart grid is distributed information network, which will measure the status of the entire or part of the power grid, and will control the energy flow according to defined policies. With such significant changes required in architecture and operations, the need of a practical and laboratory-based research environment becomes valuable. In this work we have designed a laboratory-based practical smart grid research testbed, called *SmartGridLab*. It is an efficient smart grid testbed designed to help the research community analyze their designs and protocols for solving the open problems. This will boost the smart grid researchers to develop, analyze, and compare different protocols and designs conveniently and efficiently. *SmartGridLab* consists of the following major components: (1) intelligent power switch for power routing, (2) power supplies (main supply and renewable energy supply), (3) energy demanders (e.g., appliance), and (4) an information sharing network containing power meters. Through detailed experiments we have validated the usage of our designed testbed for conducting experiments on currently open research problems in smart grid.

8.1 Introduction

The smart grid is the vision of a more efficient, dynamic, distributed electricity grid architecture. It is supposed to deliver electricity from a distributed source of suppliers to the consumers, utilizing digital technology with two-way communications in order to control the energy distribution in a smarter way. This will involve self-operating the appliances at the consumers' side for saving energy, reducing cost, and increasing reliability and robustness. A state-of-the-art smart grid design is necessary to solve the foreseen energy crisis in many countries. In the United States and also in other parts of the world, the demand of energy is growing faster than the supply of energy. This has often caused peak energy demand to approach the energy grid capacity, thus causing frequent blackout. The literature [1] shows the historical outage statistics of the United States from 1991 to 2005. Figure 8.1 reveals the trend of faster growth of energy demand with respect to energy supply in the United States from 1950 to 2008. In the existing power grid,

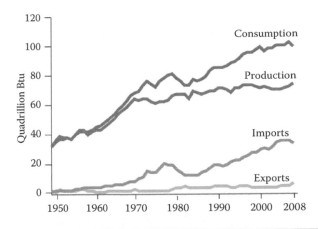

Figure 8.1 Trend of energy production and consumption in the United States. (Source: Energy Information Administration, Energy Perspectives, Figure 1 (June 2009).)

the basic principle of transferring energy from a power plant to a large number of users cannot often meet the increasing demand. There have been five massive blackouts over the past 40 years, three of which have occurred in the past 9 years [2]. The Northeast blackout of 2003 is one of the worst, and it shows that the traditional electricity grid's self-healing capability is not robust enough. Since the traditional grid is a centralized system, if some critical link is broken, the components that are connected with this link will also be broken, and may lead to cascaded failure.

Because of these problems, a trend is to seamlessly integrate the sources of renewable energy supply and allow distributed power generation. This will not only reduce peak load, but will also reduce important factors like CO_2 emission, greenhouse effect, and energy consumption. One more advantage of such integration is quicker recovery of communities in disaster scenarios, as they need not rely on the quick recovery of the main power grid. The sources of renewable energy are very promising in enabling such design. In 2008, 11.8% of electricity was generated from renewable energy sources in California [3]. According to California's Renewable Energy Programs, by 2020, 33% of electricity will be generated from renewable energy resources. In Europe, 8.5% electricity was generated from renewable sources in 2005, and their goal is to increase this figure to 20% by 2020 [4].

A grid architecture connecting distributed energy sources and users (such as in Figure 8.3) is more suitable for smart grid design than the centralized grid architecture (as shown in Figure 8.2). Such a grid has the useful properties to interconnect distributed sources of energy supply and varied consumers, and to offer better disruption resilience. In a mesh grid, even if some links are broken, it will not affect the whole power grid.

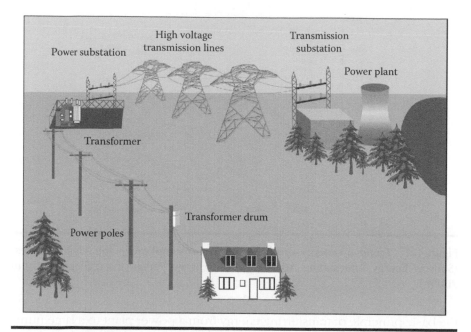

Figure 8.2 Traditional centralized electricity grid. (Source: http://science. howstuffworks.com/environmental/energy/power.htm.)

This greatly reduces the possibility of large-scale blackout. A traditional grid is only a one-way energy broadcasting network. However, in the future more renewable energy sources will be used, and the power grid will have two-way energy transmission to support users to use energy, as well as upload their extra energy to the grid and share it with others.

In the smart grid research community, simulation is widely used. Karnouskos and de Holanda [5] have designed a simulator based on software agents that attempts to create the dynamic behavior of a smart grid city. [6] Integrid is a grid simulation laboratory, working on tests and validation of computer simulations with system dynamics when renewable generation sources and other forms of distributed generation and loads are integrated into the electric power grid. But simulated environments lack real scenarios and platforms to conduct experimental research in the laboratory environment. In order to develop, analyze, and compare different designs efficiently and solve the related open problems, a laboratory-based smart grid testbed is more effective. Therefore, to foster the ecosystem of smart grid research, we have developed a smart grid testbed for the laboratory research environment. This testbed will allow researchers and educators to effectively study and teach smart grid technology. It has the following main components: (1) intelligent power switch, (2) power supply (with main supply as well as renewable energy sources), (3) energy

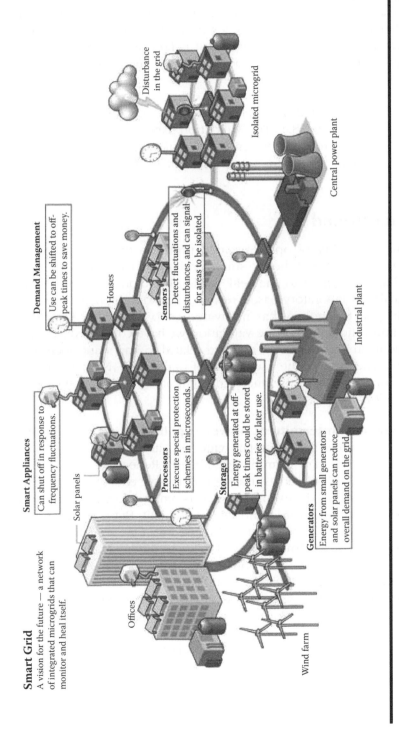

Smart Grid

A vision for the future — a network of integrated microgrids that can monitor and heal itself.

Smart Appliances

Can shut off in response to frequency fluctuations.

Demand Management

Use can be shifted to off-peak times to save money.

Solar panels

Offices

Houses

Wind farm

Sensors

Detect fluctuations and disturbances, and can signal for areas to be isolated.

Processors

Execute special protection schemes in microseconds.

Storage

Energy generated at off-peak times could be stored in batteries for later use.

Generators

Energy from small generators and solar panels can reduce overall demand on the grid.

Disturbance in the grid

Isolated microgrid

Central power plant

Industrial plant

Figure 8.3   Envisioned smart grid architecture. (Source: http://www.infowars.com/smart-grid-the-cloud-why-are-they-linked/.)

demander (e.g., appliances), and (4) information network with power meters. The intelligent power switch can dynamically configure the route of power from a set of energy supplies to a set of users. The power meter can measure the amount of energy that is flowing through the lines. The information network processes the status of the network and controls the energy distribution.

This work is organized as follows. The background for our work is discussed in Section 8.2. The detailed design of the *SmartGridLab* testbed is pesented in Section 8.4. The validation experiments on *SmartGridLab* for smart grid research are presented in Section 8.5. Finally, in Section 8.6 we conclude with discussion on future work.

8.2 Background

There are a number of open research problems that need to be solved for achieving a state-of-the-art smart grid architecture (discussed in reference [2]), applicable to the real world. Therefore, the requirement of a practical and laboratory-based research and development environment for smart grid technology has motivated us to design the *SmartGridLab* testbed and validate it. For the validation we have conducted a number of experiments in *SmartGridLab* that are related to important open research problems. These show the usefulness of our testbed. Now we briefly discuss the open problems in smart grid that have influenced the design of *SmartGridLab*.

Management of intermittent power supplies: Various renewable energy sources such as solar, wind, etc., are being used in smart grid architecture. For example, California's Renewable Energy Program is going to increase its usage of renewable energy resources to 33% by 2020. The inclusion of various renewable energy sources will enrich the smart grid and its applicability. But there are a number of challenges associated with their use. One challenge is how to manage the intermittent availability of different energy supplies in the smart grid, so that consumers can get continuous energy supply according to their demand. The size and capacity of renewable energy sources are relatively smaller, so consumers can use them in a localized part of the grid. But the small-capacity supplies can not only provide energy to these consumers, but also be viewed as a virtual power plant to provide energy back to the grid. Control and communication algorithms are needed to make the network of renewable energy sources more efficient and reliable. This requires the design of a dynamic and adaptive solution of energy supply from hybrid energy sources to varied consumers.

Price-driven real-time demand response: Demand response (DR) is the ability of consumer components to dynamically change the electricity loads of demand. The change can be according to price signal, which may reflect the total demand of the user side. DR is one of the important capabilities to enable the smart grid. To enable DR, several techniques are needed. Users need smart meters to report their current or predicted future energy consumptions, and appliances need to adjust their consumption behaviors and schedules accordingly. Sensing of renewable energy production rates and the changing prices is also an important issue. Protocols for sending price signals and algorithms to control appliance behaviors need to be developed. The varied sources of energy and the consumers of energy will be distributed in the entire grid. This will lead to the challenge in designing how energy will be distributed from multiple suppliers to multiple consumers in an efficient and dynamic way.

Disruption resilience with self-healing: A traditional power grid mostly doesn't have the self-healing capability. The failure of some links in the power line may result in loss of electricity to consumers on a large scale. However, a smart grid will have a monitoring system that will learn the status of the whole power grid. The goal is to dynamically optimize the performance and robustness of the system, and to quickly react to disturbances in order to minimize the impact (e.g., cascaded failure). Microgrids and virtual power plants will also potentially provide such disruption resilience capability to the smart grid.

Communication architecture: The difference between a traditional grid and a smart grid is that the smart grid relies more on communication between consumers, suppliers, smart devices, and applications. To support home area networks (HANs), building area networks (BANs), and industrial area networks (IANs) in a smart grid, several kinds of networks have to be considered for suitability. High-speed networks such as WiFi, power line communication, and lower data rate communication (such as 802.15.4) could be suitable for different networks. All these networks have to be integrated into one larger network to provide efficient data flow and control flow. To make these networks suitable for the smart grid, new protocols and algorithms are needed.

Dynamic pricing: Because of the distributed nature of energy sources, the consumers in a smart grid will have options to get energy supply from their choice of source, according to the need of quality of provided energy and the price of energy. This requires design of choice-driven power draw and consumption from available sources.

Reduction in energy loss: The power distribution lines are often lossy with resistance and reactance. Therefore, the power flow in a large-scale smart grid needs to take care of energy loss, so that most of the power can be delivered with minimum loss. This requires design of protocols for the power flow methodology through the grid.

Scheduling of power consumption to constrain peak load: The future smart grid will consist of smart environments (smart homes, smart workplaces, etc.) that will be equipped with smart electrical appliances. This will enable proper integration of smart grid architecture. The typical view of a future smart environment is shown in Figure 8.4. In the near future, the electricity prices will increase/decrease as the peak load increases/decreases. The future appliances in smart homes will be smart and will have the capability to delay their operation for a reasonable period (e.g., 10 seconds) to minimize the peak demand, thus minimizing the cost. The goal

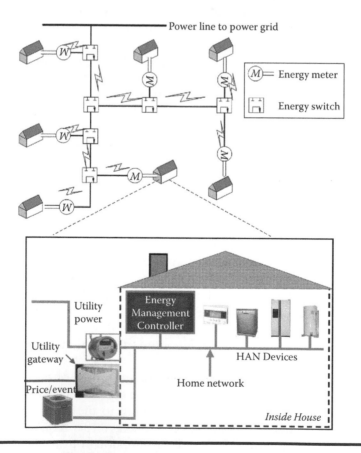

Figure 8.4 Power distribution and control system for future smart environments (e.g., smart home) in smart grid.

of the next version of *SmartGridLab* is to build a sensor network of power meters and controllers, then design a scheduling algorithm to schedule appliances to minimize the power peak load. This peak load control methodology equivalently applies to the greater power grid, where each home/community energy consumption is controlled in order to limit the peak load of the microgrid or that component of the greater smart grid.

These challenges and opportunities for smart grid design have influenced the development of *SmartGridLab*.

8.3 Related Work

In the smart grid research community, simulation is widely used. Karnouskos and de Holanda [5] have designed a simulator based on software agents that attempts to create the dynamic behavior of a smart grid city. Integrid [6] is a grid simulation laboratory, working on tests and validation of computer simulations with system dynamics when renewable generation sources and other forms of distributed generation and loads are integrated into the electric power grid.

In He et al., [7] a concept of a future power grid is presented; it proposes an innovative electric power architecture. This architecture learns from both the Internet and microgrid. The main component of this architecture is an intelligent power switch (IPS) with an energy generator and energy storage on it. This design is suitable for renewable energy and easy to connect with the current power system.

Microgrids are discussed in Driesen and Katiraei [8] and in Robert and Lasseter [9]. A micogrid is a cluster of generators and loads. It is a new concept in which a cluster of loads and distributed generation (DG) systems operates to improve the reliability and quality of the power system in a controlled manner. The generator could be a renewable energy generator such as a wind turbine or solar panel. It can improve the quality and reliability of the power grid. A microgrid has the ability to disconnect from the main grid when there is a disruption in the main grid. After disconnection, the island grid can use the energy source inside it to supply power. When disruption stops, it can reconnect to the main grid. For customers, microgrids provide the need for power in a reliable way. For the whole system, microgrids are dispatchable cells that can respond to the signals from the system operator very fast. Information technology achievements along with new DG systems with intelligent control systems allow system operators and microgrid operators to interact in an optimal manner.

There are several works discussing power meter design. MIT's plug node [10] uses a current transformer as the current sensor, and an ADC to sample this sensor. UC Berkeley's ACme [11] is an IP-based wireless AC energy meter. It uses ADE7753 as a current sensor.

Various communication methods have been used in the smart grid. TUNet, the Tantalus Utility Network, is an end-to-end WAN/LAN/HAN communications system that operates with both RF and IP-based networks, including fiber, WiFi, WiMAX, and GPRS/cellular, either individually or in combination [12]. GridComm [13] is based on the Tropos wireless broadband mesh network system. GE has announced use of WiMAX in a grid pilot program for Michigan utility Consumer Energy [14].

To minimize power loss in transmission and distribution, some minimum cost-flow algorithms were developed in Kashem et al. [15] and Ramesh et al. [16].

8.4 Smart Grid Testbed Design

Now we discuss the design of *SmartGridLab* in detail. The *SmartGridLab* testbed consists of four main components: (1) intelligent power switch (IPS), (2) energy supplier (main supply, and renewable energy sources like solar panel and wind turbine), (3) energy demander (e.g., appliances), and (4) an information network containing power meters.

In the *SmartGridLab* testbed, there are two networks that coexist and cowork with each other. One is the power distribution network with energy flow, and the other one is the information network with flow of sensing and control data. The information network can collect the status of the power network and can also control it according to some defined policies. Both the power and information networks have two-way flow. We have developed an intelligent power switch (IPS), which acts as a power router in the power grid. An IPS is a device that can reroute the power flow from one input to another. It is equipped with a microcontroller and wireless communication components within it. The detail of an IPS is presented in Section 8.4.3. A power meter is used in this testbed to measure the power flow through lines. The detailed design of a power meter is presented in Section 8.4.4.

8.4.1 Architecture of Power Network

Figure 8.5 illustrates the overview of a power grid using an IPS (as discussed in He et al. [7]). IPS can be connected to energy sources (including renewable energy, main source of power grid, etc.), smart appliance, energy storage, power meter, and also to more IPSs. So IPS is a critically important component for achieving smart grid architecture. Based on this feature, the grid that contains an IPS can be configured into various different topologies. An IPS can be connected as a mesh, tree, or ring, and

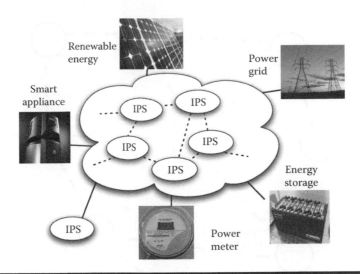

Figure 8.5   Power grid overview.

the configuration can also be changed as desired. Figure 8.5 shows a distributed structure of a power grid. The distributed power suppliers and consumers are connected to the cloud of the IPS. By connecting to the IPS, a new component can easily be added into the power grid. In such a power network no centralized control is needed. Rather, it is like a peer-to-peer network. The IPS can be connected to the current power grid system. It can also act like a microgrid. It can group the devices that are connected to it and can isolate from the main power grid if any disruption is detected.

8.4.2 Architecture of Information Network

In the smart grid, a two-way data communication network will allow information exchange. A variety of communication media could be used in the smart grid information network. We have used the 802.15.4 wireless network (configured as a wireless mesh network) in this testbed to emulate the network. Although WiFi or wired media can also be used for communication, the 802.15.4 is low power and is more flexible for a testbed. It can be configured into any topology. This information network is a kind of sensor network, and the power grid is the object it senses.

This information network can be configured into a centralized network or a distributed network. For the centralized configuration, power meters can send their data to an energy management center (EMC). EMCs can compute the whole power grid status and send out a control signal. In distributed configuration, each microcontroller on the IPS will compute its own status based on the information it received from another IPS and

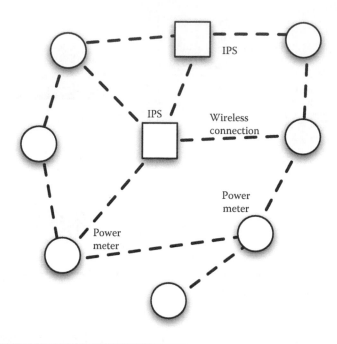

Figure 8.6   Wireless information network.

power meter. Each IPS and power meter can communicate with each other and exchange their status.

8.4.3 IPS Design

The IPS is the critically important component for achieving distributed and scalable architecture, and it needs intelligence to efficiently control the interconnection of components. The purpose of an IPS is to switch power from one port to another. So any component connected with an IPS will be connected with the others. Through an IPS, at the same time multiple pairs of ports can be connected together. That means if there are two supplies, A and B, and two consumers, C and D, A can provide power to C, and B can provide power to D in parallel. To achieve these goals, the IPS uses the design as shown in Figure 8.6. Ports can be connected with an appliance, power supply, or another IPS. Switches on the intersection of two lines will control the connection of a pair of lines. Taking S6 as an example, it is on the intersection of ports 3 and 4. If it is closed, the two ports will be connected together; otherwise, there is no connection between them.

This design in Figure 8.8 is of a six-port IPS. Fifteen switches are used to control six ports. If more ports are to be used, there should be more switches. Assuming that the number of ports are N_p and the number of switches are N_s, then for individual IPS:

$$N_s = \frac{N_p \times (N_p - 1)}{2} \tag{8.1}$$

So if a lot of ports are needed in some application, we can use two or more IPSs and configure them in a cascade connection. The configuration of connection can be of three types: multiple suppliers to single consumer, single supplier to multiple consumers, and parallel connection. For example suppose the requirement is port 1 only supplies energy to port 3, and port 2 only supplies energy to port 4. To achieve this, the IPS closes S11 to connect port 1 and 3, then closes S7 to connect ports 2 and 4.

The hardware of IPS is shown in Figures 8.7 and 8.9. This is a six-port IPS; each outlet can be connected to either a power supply, an appliance, an energy storage, or another IPS. In this design, a TelosW platform [17] is used as a controller of the IPS. It can send out a control command to shift register. Two shift registers are used in this design; each of them has eight outputs, and each output can control one of these switches. As the power supply, it can get power directly from the power line. In this version,

Figure 8.7   Intelligent power switch.

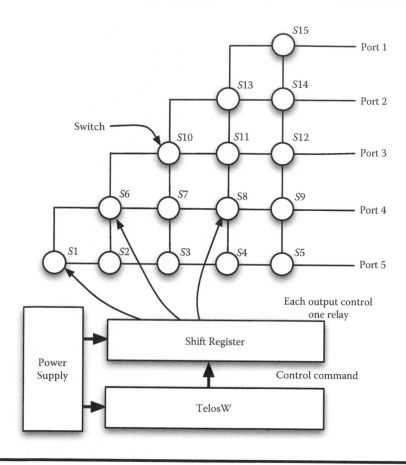

Figure 8.8 Intelligent power switch design.

the power supply should be independent from all six outlets. In the next version of the IPS, we plan to use a rechargeable battery to supply power to TelosW and the switches. It can then be charged whenever one of the six outlets is connected to the power supply. Solid-state relays S116S01 [18] are used as the power switch. The S116S01 can provide 4.0 kV isolation from input to output, and the peak off-state voltage is 400 V. By using this device, it is easier to control high-voltage AC by a low-voltage control signal.

8.4.4 Power Meter

The power meter is another important component in this testbed. It can measure how much current is flowing in the lines. This power meter is built with four components: a TelosW sensor mote, Hall effect current sensor, resistor network, and power supply. The TelosW is the controller of

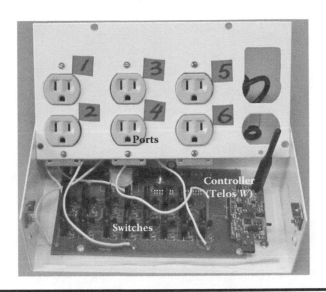

Figure 8.9   Intelligent power switch hardware with input ports.

the power meter. The Hall effect current sensor converts current value to voltage. In this design, the ACS714 [19] 5A version is used. It has 1.2 mΩ internal conductor resistance, so its energy consumption is negligible. The output of the ACS714 is linear according to current change on the test line. The output of the ACS714 can be converted to digital numbers by an analog

Figure 8.10   Power meter hardware.

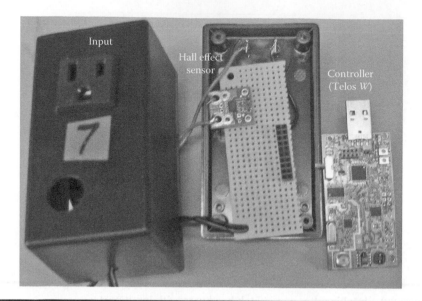

Figure 8.11   Power meter hardware with power plug.

digital convertor (ADC), so the current value can be processed by a micro-controller. However, the output voltage of the ACS714 varies from 1.5 to 3.5 V, while the ADC on the TelosW can only allow a maximum 2.5 V input. So a resistor network is needed to regulate the input below 2.5 V. It can get power supply directly from the power line, and output is a stable 5 V to the ACS714 and TelosW. The hardware of our power meter is shown in Figure 8.11 and the design shown in Figure 8.12.

We use the MSP430F1611 [20] as a microcontroller and use its own ADC to sample data from the Hall effect sensor. Considering the frequency of the MSP430F1611 (which is only 8 MHz) and the data rate of the radio, if the sampling rate is set as 0.1 mS, it cannot send out all data to the EMC. So local data processing is necessary. The flow of process is shown in Figure 8.13.

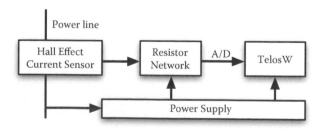

Figure 8.12   Power meter design.

Figure 8.13   Software of power meter.

First, 512 samples are taken from ADC in each 0.1 mS. After this is done, the root mean square (RMS) is computed based on these samples. In AC signal, RMS can indicate the average current. Once this is done, the final result is sent to EMC through radio communication. One problem in this flow is that for computing RMS, it will take a lot of time. However, the energy consumption of appliances will not change very quickly, so the computation of RMS can be done at a low rate. In our testbed, we take a 1-second interval between two computations of RMS.

8.4.5 Energy Supply and Energy Demanders

In this testbed, wall outlet power and renewable energy sources can be used as power supply. They can be connected to IPS to provide energy to the rest of the power network. Figure 8.14 shows two micro renewable energy generators. They are small enough to be used in laboratory experiments. As energy demanders (i.e., consumer) we have used lamps, computers, and other appliances. We have also designed a smart appliance (Figure 8.15) that can intelligently control the energy usage according to price signal of supplied power.

Figure 8.14   Renewable energy source.

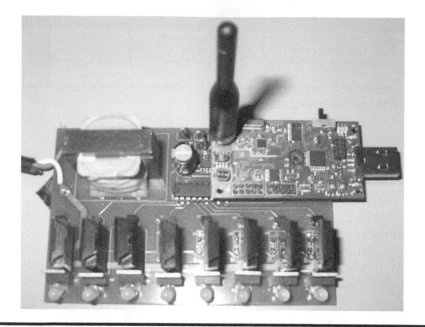

Figure 8.15 Smart appliance design.

Figure 8.16 Experiment setup.

8.5 Testbed Validation

Various research problems can be studied using the *SmartGridLab* testbed. In this section, we have demonstrated smart grid system experiments with *SmartGridLab*. Six IPSs are used and interconnected. In our experiment we configure these IPSs into a mesh power network; however, other kinds of topology could also be formed by IPS. On the connection of each IPS there is a power meter to measure power flow between them. Two power supplies, P1 and P2 (shown in figures), provide power to the network. They are connected to switches S1 and S4. We use three lamps to simulate the energy consumer (Figure 8.18). They are indicated as A1, A2, and A3 in the figure. A1 and A2 are connected with S5, while A3 is connected with S6. In the figures, the real power meter reading is shown.

8.5.1 Real-Time Demand Response

We have conducted experiments related to two demand response strategies: (1) reliable energy supply with multiple intermittent sources, and (2) price-driven demand response with multiple flows.

8.5.1.1 Management of Intermittent Power Supplies

One issue with using varied sources of renewable energy in the smart grid is intermittent availability. The output of some energy supplies may not be stable. However, if the system used multiple such supplies to form a virtual power plant, the whole output can become better and more stable. In this experiment, we emulate two renewable energy sources with intermittence on each of them, but the intermittence does not happen at the same time, and the appliance still gets a continuous power supply. In Figure 8.17, the first two plots are of two power supplies with intermittence. We emulate them by turning on and off the connection of the switch connected to the power supply. From these two plots, it can be observed that none of them has continuous output. The third plot is the energy consumption of the appliance (A1, 100 W lamp is used), and it can be seen that the energy supply has been stable.

8.5.1.2 Price-Driven Demand Response with Multiple Flows

Demand response is an important issue with smart grids. Users will change their energy consumption level according to the price information. The price can increase if the demand becomes higher. In the smart grid, there could be many energy suppliers. Different sources of energy supply may have different prices, based on the energy quality, and different users may have different preferences. Some users may want the price to be lowest, while other users may take energy quality as their first

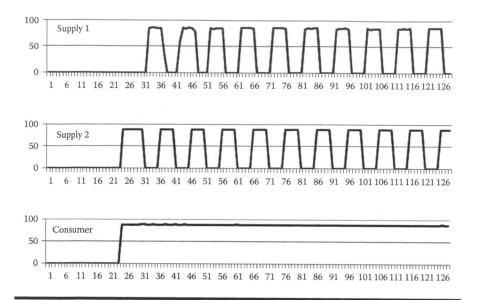

Figure 8.17 Management of intermittent power supplies: A1 getting continuous power from two intermittent sources of power supply.

consideration. So in the smart grid, multiple power flows may exist in the network at the same time. Our experiment is to show multiple power flows can coexist by using IPS. In Figure 8.18, A3 is a 40 W lamp that gets energy from supply P1 through path P1→S1→M1→S2→M6→S6→A3, while A1 is a 100 W lamp that gets energy from supply P2 with path P2→S4→M7→S6→M8→S5→A1. The two paths can coexist in this network according to the reading of the power meter. Even though have an intersection in S6, they will not affect each other.

8.5.2 Disruption Resilience with Self-Healing

Disruption resilience is one of the smart grid's key features. The link from supply to consumer may be broken at some point. The smart grid should have the ability to switch paths from the broken link to another one. In this experiment, appliance A2 and supply P1 are used. First, A2 is connected with P1 through path P1→S1→M1→S2→M6→S6→M8→S5→A2. The link between S2 and M6 is broken. Then the path from P1 to A2 is switched from the original path to a new path: P1→S1→M2→S3→M5→S5→A2. This shows the disruption resilience capability of *SmartGridLab*.

8.5.3 Flow Balance Using Multiple Paths

In the power grid, the load of each line is limited. So it may require balancing the power flow in the power grid to make sure none of the load

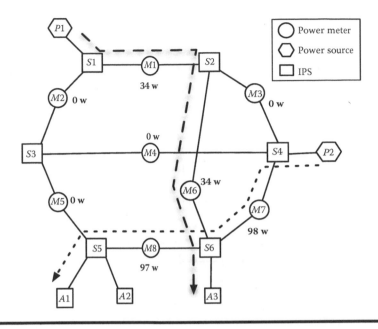

Figure 8.18 Multiple flows: A1 (100 watt) gets energy flow from P2, while simultaneously A3 (40 watt) gets energy flow from P1. The real energy flow measured across lines is also shown.

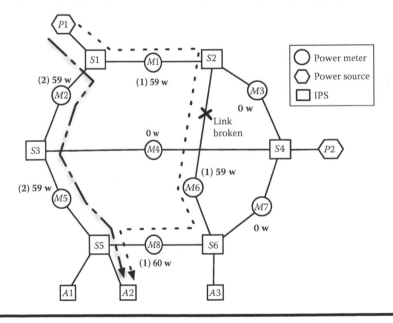

Figure 8.19 Self-healing: A2 (60 watt) initially gets energy flow from P1. When the link from S2 to M6 is broken, self-healing smart grid assigns a new path from P1 to A2. The real energy flow measured across lines is also shown.

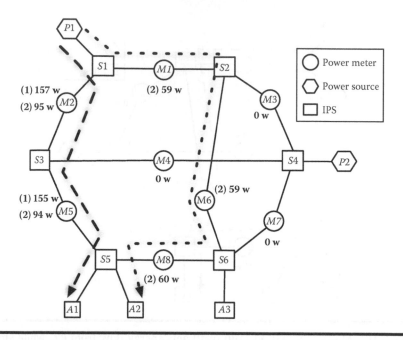

Figure 8.20 Flow balance with multiple path: A1 (100 watt) already getting energy flow from P1. Then for A2 to get energy flow from P1, the smart grid assigns another path for maintaining balance in energy flow through lines. The real energy flow measured across lines is also shown. Power meter readings shown: (1) is before flow balance, (2) is after flow balance.

is higher than the limit of the electric line. The mesh power grid platform can achieve the balance by setting the configuration of different switches in the network. In this experiment, A1 (100 W) and A2 (60 W) are connected into the network, and they get energy from P1 (Figure 8.20). However, all flow is coming from path P1→S1→M2→S3→M5→S5, and the other part of the network has no flow. Assuming that each line's limit becomes 100 watt, this path is overloaded. To meet the limit, IPS still connects A1 to P1 with the same path as before, but switch A2 to another path: P1→S1→M1→S2→M6→S6→M8→S5→A2. Figure 8.20 shows the flow (1) before balance and (2) after balance.

8.5.4 Power Meter

To validate the power meter, we set up two experiments. In the first experiment, we use three lamps (130 V, 40 W; 120 V, 60 W; 120 V, 100 W) as load. We turned them on one by one and got readings from the power meter. From Figure 8.21, we see that our power meter precisely reflected the changing power consumption. Also, Figure 8.22 shows that the power meter reading was fairly stable and accurate through time.

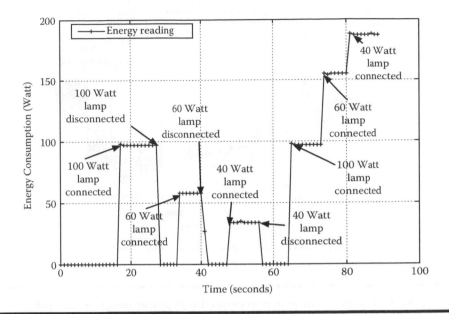

Figure 8.21 Validation of power meter.

In the second experiment, a laptop has been connected with the power meter. The energy consumption is measured for more than an hour, as shown in Figure 8.23. We first charged the battery from 85 to 100%. In the middle of this process, we run a graphic editor software, so there is a spike at 1,000 seconds. After charging the battery, we made the computer go to

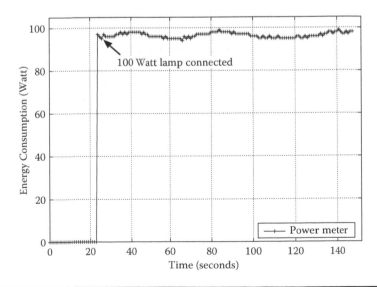

Figure 8.22 Stability of power meter reading.

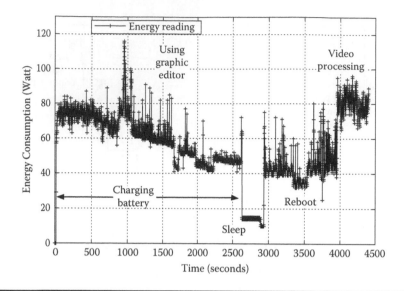

Figure 8.23 **Energy consumption (measured with power meter) of an Apple Mac-Book during different operations.**

sleep mode; therefore energy consumption drops to about 15 W. In the last part of this figure, we turn on the video processor software, which uses about 80 W.

8.6 Conclusion and Future Work

This chapter presents *SmartGridLab*, a laboratory environment testbed for smart grid research. In this testbed, the main components are IPS (for power routing), different sources of power supply, energy demander, and power meter. IPS enables the power grid to be configured as any kind of topology for power distribution, and the power meter senses the energy flow on each line and reports the data. The information network is codesigned with the power network. *SmartGridLab* can significantly help researchers to analyze and compare various algorithms and protocols. Several experiments are performed to show the applicability of this testbed for the research community in smart grid. Currently the IPS has to be plugged in to another power supply. We plan to replace it with a rechargeable battery. The power meter cannot measure active and reactive energy consumption yet. In the next effort we plan to add this feature.

References

1. M. Amin and J. Stringer. The electric power grid: Today and tomorrow. *MRS Bulletin*, Warrendale, PA: 33(4), 399–407 (2008).

2. The smart grid: An introduction. In *A Report from the Department of Energy's report from DOE's Office of Electricity* (2008).

3. *California's Renewable Energy Programs.* http://www.energy.ca.gov/renewables/index.html

4. *20% of Renewable Energy by 2020.* http://www.our-energy.com/videos/eu_20_percent_of_renewable_energy_by%

5. S. Karnouskos and T. N. de Holanda. Simulation of a smart grid city with software agents. (2009). Third UKSim European Symposium on Computer Modeling and Simulation (EMS '09), Athens, Greece.

6. InteGrid: Grid Simulation Laboratory. http://www.integridlab.com/.

7. M. He, E. Reutzel, X. Jiang, R. Katz, S. Sanders, D. Culler, and K. Lutz. An architecture for local energy generation, distribution, and sharing. In *IEEE Energy 2030, Atlanta, Georgia* (November 2008).

8. J. Driesen and F. Katiraei. Design for distributed energy resources. In *Power and Energy Magazine, IEEE, 6(3)*, 30–40 (2008).

9. P. P. Robert and H. Lasseter. Microgrid: A conceptual solution. In *PESC04*, Aachen, Germany, June 20–25, 2004.

10. J. Lifton, M. Feldmeier, Y. Ono, C. Lewis, and J. A. Paradiso. 2007 A platform for ubiquitous sensor deployment in occupational and domestic environments. In *IPSN* (April 25–27, 2007).

11. X. Jiang, S. Dawson-Haggerty, P. Dutta, and D. Culler. 2010 Design and implementation of a high-fidelity ac metering network. In *IPSN* (April 15–18, 2009).

12. Press release: Morristown hits grand slam with fiber-based tantalus smart grid network. *Department of Energy* (April 21, 2010).

13. Press release. 2010: Glendale water and power selects tropos gridcom for smart grid initiative. *Department of Energy* (April 14, 2010).

14. Change in the smart grid landscape? CISCO, GE put some muscle behind wimax. *Department of Energy* (March 26, 2010).

15. M. Kashem, D. L. M. Negnevitsky, and G. Ledwich. Distributed generation for minimization of power losses in distribution systems. In *IEEE Power Engineering Society General Meeting*, Montreal, Canada, pp. 8–14 (June 2006).

16. L. Ramesh, S. P. Chowdhury, S. Chowdhury, A. A. Natarajan, and C. T. Gaunt. 2009. Minimization of power loss in distribution networks by different techniques. In *International Journal of Electrical Power and Energy Systems Engineering* (2009).

17. L. Gang, D. Debraj, X. Mingsen, and S. Wen-Zhan. 2010. Telosw: Enabling ultra-low power wake-on sensor network. In *INSS 2010* (June 2010).

18. *S116S01 Series Datasheet, Sharp.* http://sharp-world.com/products/device/lineup/data/pdf/datasheet/s116s01_e%.pdf

19. *ACS714 Datasheet, Allegro.* http://www.allegromicro.com/en/Products/Part_Numbers/0714/0714.pdf

20. *MSP430 Datasheet, Texas Instruments.* http://focus.ti.com/mcu/docs/mcuprodoverview.tsp?sectionId=95&tabId=140&f%

Chapter 9

Deterministic Ethernet Synchronism with IEEE 1588 Base System for Synchrophasor in Smart Grid and Integration in IEC 61850 Standard

Víctor Pallarés-López, A. Moreno-Muñoz,
M. Gonzalez-Redondo, R. Real-Calvo,
I. M. Moreno-Garcia and Juan José Gonzalez
de la Rosa

Contents

At present, phasor measurement units (PMUs) are the most widely used synchronized measurement technology (SMT)-based device for power system applications. The major advantages of using an SMT are that all measurement signals are attached with a high-accuracy time stamp; this will facilitate the transition from the conventional SCADA-based measurement system to a more intelligent measurement system that utilizes synchronized measurements from geographically distant locations. This will help in spreading the smart grid conception. We propose a new synchronized technique. A precision time protocol (PTP)-based global system has been defined to provide a synchronized substation for phasor measurements. It implements the precision time protocol to perform time stamping for these innorative electronic designs (IEDs). For this objective, we have developed experimental procedures to the standards involved.

9.1 Introduction

As demands on the grid change, and the need for sophisticated information infrastructure increases, it will become more important than ever to ensure that events are accurately recorded and time-stamped consistently. Proper time synchronization across interconnections is a very important function for many reasons. Common time synchronization will be a key for many smart grid applications. The IEEE Std 1588[TM] will be a key element to achieve that synchronization. This standard is available to achieve highly accurate synchronization over a communication network. Many applications related to smart grid require time synchronization.

Synchrophasor measurements are key information needed by system operators to assess the status of the power grid. Using data from phasor measurement units (PMUs), received by phasor data concentrators (PDCs), grid operators will be able to have better visibility of power grid operations and respond to grid disturbances earlier to prevent major blackouts [18].

Two standards are related to communications of phasor measurement unit (PMU) data and information. IEEE C37.118 was published in 2005 for PMUs. IEC 61850 has been substantially developed for substations but is seen as a key standard for all field equipment operating under both real-time and non-real-time applications. The use of IEC 61850 for wide area communication is discussed in IEC 61850-90-1 in the context of communication between substations; it is only a small step to use it as well for transmission of PMU data. The models for PMU data need to be defined in IEC 61850. This work seeks integration of study with experimental tests [19].

At present, synchronized measurements based on an accurate time reference, e.g., global positioning system (GPS), provide the missing link, now allowing more efficient use of phasor data [1]. These phasor meters are very geographically dispersed through wide areas and still capture electrical waveforms in a synchronized way with a precision up to the microsecond range. The synchronization requirements are very close to the ones imposed to systems working with a unique clock.

Wide area measurement systems use the technologies of synchronized phasor measurements and modern communications, monitoring and analyzing the current operation status of a wide area power system, and serving the real-time controls and operations in power systems [2].

For improving reliability of protective relay in the digital and analog protection testing, particular attention has been paid to the use of real failure data, recorded using a digital fault recorder (DFR) or a phasor measurement unit (PMU) to test the performance of the protection device in an effective and practical way [3].

In general, these applications include power plant automation systems, substation automation systems, programmable logic controllers (PLCs), intelligent electronic devices (IEDs), sequence of event recorders, digital fault recorders, intelligent protective relay devices, energy management systems (EMS), supervisory control, and data acquisition (SCADA) systems, and plant control systems.

We propose a new synchronized technique for smart grid. A PTP-based global system has been defined and an experimental PTP-based system has been developed to provide synchronized substations for phasor measurements. It implements the precision time protocol (PTP) to perform time stamping for these IEDs.

Both approaches present obvious drawbacks. First, all the remote units need to be located in points from where at least four GPS satellites can be seen at every moment. This requirement is not always technically feasible. The second technical alternative, based on optical fiber lines having to be deployed through vast geographical areas, is well suited for industrial environments, but it turns into a very expensive option in other cases.

For example, synchrophasor standard IEEE C37.118 [4] imposes critical synchronism requirements. To keep the total vector error (TVE) level 0 (highest) below 1% threshold, the highest phasor angle error allowed is 0.57°, on a 50 Hz nominal frequency for an electrical network (all data from now on are referenced to 50 Hz nominal frequency networks). A time error of 10 μs corresponds to a phase error of 0.18°. Furthermore, our technical proposal integrates a variety of features in order to reduce to a minimum synchronism errors in the signal sampling and conversion process [5].

9.2 Synchrophasor Standards

Synchrophasor measurements are key information needed by system operators to assess the status of the power grid. Using data from phasor measurement units (PMUs), received by phasor data concentrators (PDCs), grid operators will be able to have better visibility of power grid operations and respond to grid disturbances earlier to prevent major blackouts. The current primary standard for the communications of PMU and PDC data and information is IEEE Standard C37.118, which was published in 2005. This standard also includes requirements for the measurement and determination of phasor values [18].

A synchrophasor is the same phasor, except that the measurement is relative to a sinusoid at the nominal system frequency synchronized to UTC time. Any frequency deviation from nominal causes a rotation of the synchrophasor phase angle. Measurement requires a specialized phasor measurement unit (PMU) that can sample the sinusoidal waveforms and estimate a synchrophasor equivalent. A PMU can be a stand-alone device or a function within another device.

However, the PMU can only observe waveforms, sample them over an appropriate interval (window), and employ suitable methods to estimate phasor equivalents of the input waveforms [20].

9.2.1 IEEE Standard 1344

The first synchrophasor standard, IEEE 1344 [6], was completed in 1995. It was created to introduce synchrophasors to the power industry and set basic concepts for the measurement and methods for data handling. It introduced a phasor measurement unit (PMU), which is a device or a function that estimates a synchrophasor equivalent for an AC waveform. This standard defined synchronized measurement using precise timing sources, and formalized an extension for inter-range instrumentation groupcode (IRIG-B) [7], which has been adopted by industry [20].

9.2.2 IEEE Standard C37.118-2005

The current synchrophasor standard, IEEE C37.118 [4], completed in 2005, was a major upgrade of the original standard. It included a measurement performance metric as well as improvements in the data communication specification. The purpose for both standards was to provide measurement and data exchange requirements to facilitate development of compatible measurement equipment and applications. In the first major improvement, C37.118 added a method for evaluating a PMU measurement and requirements for steady-state measurement. Total vector error (TVE) compares both magnitude and phase of the PMU phasor estimate with the theoretical phasor equivalent signal for the same instant of time. Also, this standard refers to the possible future use of PTP as a precise timing source.

Figure 9.1 shows how tasks are scheduled for every second interval. Voltage and current phasor estimations, time stamp, measured frequency,

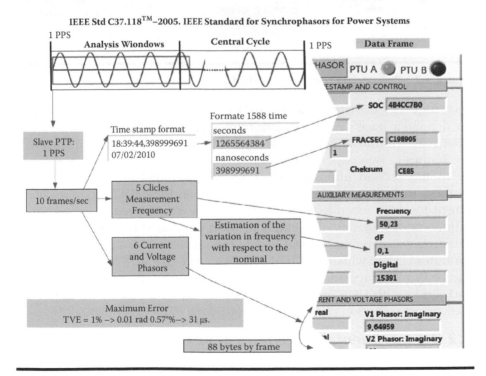

Figure 9.1   Data frame definition.

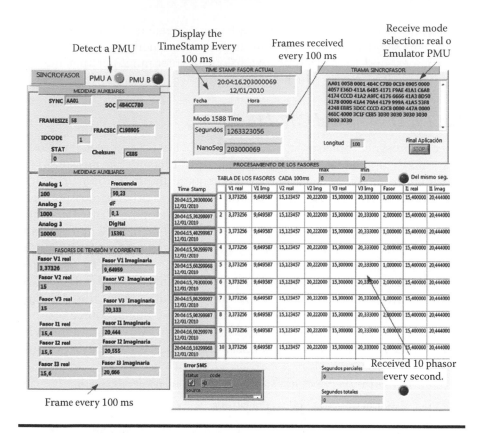

Figure 9.2   Phasor data concentrator (PDC).

and frequency deviation with regard to analysis window data are informed on a regular basis (25 or 50 data frames per second sending rates). The information is transmitted by a channel UDP unicast. A central team receives the frames of two experimental IEDs.

In the second improvement, C37.118 expanded the communication method to include higher-order collection, and the data reporting characteristics were extended. The basic status was improved to include indications of data quality. PMU identification was added to all messages. The structure was extended to enable data transmission from a phasor data concentrator (PDC), which included data from several PMUs, as can be seen in Figure 9.2. This application receives the frames that are sent by each PMU and works with UDP unicast protocol. With this application we can verify the correct reception of all frames sent by the PMU experiment.

9.2.3 Experimental System for IEEE C37.118

For this type of essay we use one NI PCI-1588 working as the slave. This card incorporates one real-time-system integration (RTSI) [21].

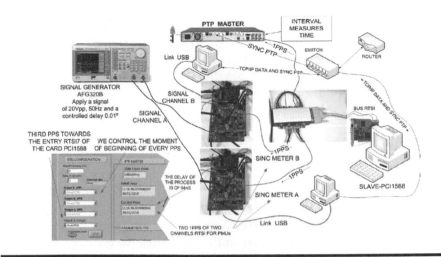

Figure 9.3   IED architecture.

All modules, connected with an real time system integration (RTSI) cable, receive the same RTSI signals. This feature makes the RTSI lines convenient in situations where you want, for example, to start an acquisition on several devices at the same time, because all modules will receive the same signal. To use RTSI signals to communicate with other NI PCI modules, you need RTSI cabling to connect the signals between the boards.

In this project we have chosen to use them as a source of synchronism for our external PMUs. For the external connection we have chosen for a cable and a connector type DB-25 connected to one of the inputs of a free slot PCI. In Figure 9.3 it is possible to see as it spreads to synchronize the PMU. In the box of connection 8 signs, RTSI is available.

All the RTSI signals of exit are generated from the oscillator IEEE 1588. The phase errors between the RTSI pulses are minimal, and some 5 ns. With this method we guarantee the simultaneity of the pulses for two different PMUs. In one of the phases of the analysis it is interesting to us to determine the errors exclusively due to the apprehension and prosecution of two PMUs.

With this method we annul the errors for different PPS and for the analogical earnings, two sinusoidal signs of 20 Vpp and 50 Hz proceeding from a professional generator. In these conditions the errors are almost exclusively attributable to the architecture of the PMU.

9.2.4 New Synchrophasor Standards PC37.118.1 and PC37.118.2

PC37.118.1 will be the new IEEE standard for synchrophasor measurements for power systems and cover phasor, frequency, and ROCOF measurement (ROCOF is the acronym for rate of change of frequency).

PC37.118.2 will be the new IEEE standard for synchrophasor data transfer for power systems and covers only the communications. The synchrophasor communication requirements in the current PC37.118 standard do not define a complete protocol; they only specify data content and format and a minimal command set for communication.

9.3 PTP Standard

System designers must account for the limitations created by these variables because as transmission distance increases, it is more difficult to share signals between systems to keep them synchronized [15].

This trade-off between precision and distance presents a problem: to have a high precision of synchronization, you must have a clock with high frequency and accuracy, which can degrade as the distance between chassis, or nodes, increases.

You need another method of conveying the clock and trigger signals from the master node to the other slave nodes in the system. This method is called time-referenced synchronization.

At present, synchronized measurements based on an accurate time reference, e.g., global positioning system (GPS), provide the missing link, now allowing more efficient use of phasor data [7]. These phasor meters are very geographically dispersed through wide areas and still capture electrical waveforms in a synchronized way with a precision up to the microsecond range. The synchronization requirements are very close to the ones imposed to systems working with a unique clock.

Precision time protocol is a time transfer protocol defined in IEEE Std 1588. It has been developed to improve precision over current Ethernet protocols. Where the pervasive NTP can synchronize to within tens of milliseconds PTP has the ability to synchronize within tens of microseconds, offering a significant increase in precision. The protocol functions autonomously, discovering other PTP devices automatically, thus maintaining synchronization even in a dynamic environment [22].

9.3.1 IEEE 1588 v1 2002

IEEE 1588 is a packet-based protocol that you can use over the Ethernet. It defines a standard set of clock characteristics and value ranges for each characteristic. By running a distributed algorithm, called the best master clock (BMC) algorithm, each clock in the network identifies the highest-quality clock, that is, the clock with the best set of characteristics.

The highest-ranking clock, called the grandmaster clock, synchronizes all other "slave" clocks. If the grandmaster clock is removed from the network, or if its characteristics change in such a way that it is no longer the best clock, the BMC algorithm helps participating clocks automatically determine the current best clock, which becomes the new grandmaster. This algorithm offers a fault-tolerant and administrative-free way of determining the clock used as the time source for the entire network [23].

The grandmaster clock periodically issues a "sync" packet containing a time stamp of the time when the packet left the grandmaster clock. The grandmaster may also issue a follow-up packet containing the time stamp for the sync packet. The use of a separate follow-up packet allows the grandmaster to accurately time-stamp the sync packet on networks where the departure time of a packet cannot be known accurately beforehand.

Although the protocol adds significant advantages, there are several factors that prevent wide adoption of PTP over NTP in Ethernet networks. The precision timing protocol (PTP) also has limitations over wide area networks (WANs) and does not currently support encapsulation or authentication, as is available with the network time protocol (NTP). It is important to note that future versions of NTP could theoretically use the same hardware stamping techniques utilized in PTP to increase timing resolution, but would have the same requirement for supporting hardware.

PTP is designed for precision, not necessarily accuracy. This distinction is one of the factors referenced above for adoption of PTP over wide area links. A highly deterministic link may provide for precision, but not for accuracy; a nondeterministic link doesn't provide for either precision or accuracy [23].

9.3.2 PTP Grandmaster for Test

For experimental tests we use XLi IEEE 1588 grandmaster. This system is a GPS-referenced IEEE 1588 grandmaster clock and IEEE 1588 accuracy measurement system. This system contains a dedicated 1588 time-stamp processor. Operating at 100 base-Tx line speed with deep time-stamp packet buffers, the XLi grandmaster can support thousands of 1588 slaves.

Ideal for measurement purposes, the XLi grandmaster can also operate as a 1588 slave. Network elements such as hubs and switches degrade time transfer accuracy when using 1588 over the Ethernet. Switches in particular add nondeterministic latency and jitter to packet transit times from a 1588 master to 1588 slaves. As a result, the 1588 slave synchronization accuracy is degraded from that of the master.

9.3.2.1 The XLi's Time Interval/Event Time (TIET)

This feature can be used to measure PTP synchronization across timing networks. The XLi IEEE 1588 clock is characterized by the following nominal specification: frequency output accuracy, $< 2 \times 10^{-12}$; frequency/timing, Allan deviation; stability, 1×10^{-9} at 1 s, 2×10^{-10} at 1,000 s 1×10^{-12}

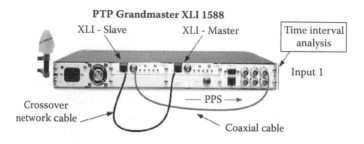

Figure 9.4   Stability of slave built on the same system.

at 1 day. The XLi comes with the standard TCVCXO oscillator described below. The stability of the following oscillators is dependent on the reference source (GPS). GPS is characterized by tracking up to 12 satellites with TRAIM; position accuracy typically < 10 m when tracking four satellites; TRAIM mask, 1 μS, 1 PPS accuracy; UTC-USNO, ±30 ns; RMS, 100 ns and peak by a PPS accuracy within 15 ns to GPS/UTC. Figure 9.4 shows a configuration to determine the stability of a slave built on the same system. The error is due exclusively to the latency of the cable itself.

9.3.2.2 Test to Determine the Stability of the Network Topology.

There are tests to determine specifically the influence of different topologies on the stability of a slave. This case is tested with the same slave in all cases. We can simulate traffic to determine the influence of the topology with different load levels.

Also, as seen in Figure 9.5, the grandmaster XLI includes a second IEEE 1588 card housed in a second slot. This card can be configured to work

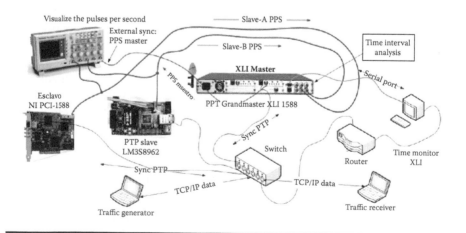

Figure 9.5   Stability for different topologies.

as a slave PTP. It is ideal for the study of the effects of different network topologies on minimizing the effect of the slave.

In this case the analysis can be extended with a second parameter as the PDV. In this case measured phase differences are estimated by the logic of the slave card in relation to the estimated difference between received synchronization packets and the time recorded in the time stamp with the master source.

This second parameter is important because it is a permanent estimate between the level of convergence achieved by the slave to master, and therefore a permanent offset estimation.

9.3.3 Experimental System for Test with PTP v1

The standard XLi IEEE 1588 grandmaster clock [1], as can be seen in Figure 9.6, provides a complete implementation of a precise time protocol (PTP) "ordinary clock" over a dedicated IEEE 1588 card. The IEEE 1588 card can be configured to operate as a PTP grandmaster or as a PTP slave.

As a PTP grandmaster, the IEEE 1588 card typically synchronizes PTP slaves on the network to international atomic time (TAI). The XLi IEEE 1588 clock derives TAI from the global positioning system (GPS). In addition, Symmetricom designed the XLi IEEE 1588 clock so the user can distribute coordinated universal time (UTC) or user-entered time over PTP.

Synchronization performance depends on several factors, including, but not limited to slave oscillator quality and PLL control [4], networking equipment, network traffic levels, and network topology. A system designer generally cannot easily modify a slave oscillator and control. However, PTP settings and network design are under the control of the system designer.

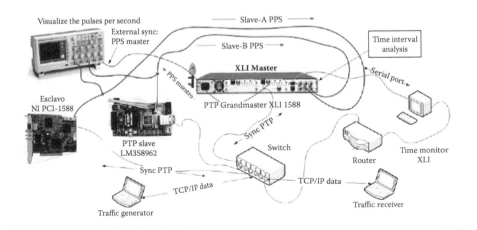

Figure 9.6 Network measurement test setup.

Through careful network design, synchronization performance of measurement systems can be maintained. Network characterization is an important step for determining the fitness for high-performance synchronization. Two parameters that aid the characterization process are packet delay variation (PDV) and slave PPS time error. PDV measures variations in the master to slave packet delay at the physical layer of the network. Measuring slave PPS time error from the hardware-generated PPS signals provides direct observation of master-slave end-to-end synchronization. Errors can be viewed using a frequency counter, oscilloscope, or grandmaster equipped with an integrated time interval measurement input XLi IEEE 1588.

9.3.4 PTP Slaves for Test

The slave is developed by Stellaris LM3S8962 evaluation board layout. The LM3S8962 microcontroller is based on the ARM®Cortex™-M3 [16] controller core operating at 50 MHz. The LM3S8962 also features hardware-assisted support for synchronized industrial networks utilizing the IEEE 1588 precision time protocol (PTP) [17]. High-precision time stamps can be achieved with the support of specialized hardware interfaces in the physical layer of the network.

The expected performance of both PTP systems (XLi-IEEE 1588 Master and two LM3S8962-PTPd slave) was tested by Symmetricom with the following conditions: synchronization was performed for 10 minutes Figure 9.7 before testing began; test durations were 2 hours; a sync interval of 2 seconds was used for all tests, and all Ethernet connections were 100 Mbps. For the switch test, a of non-1588 Ethernet traffic was present on the switch. A PTP slave continuously recovers its stability around 12.5 μs. Nevertheless, with a slave PCI-1588, the test recovers its stability around 150 ns.

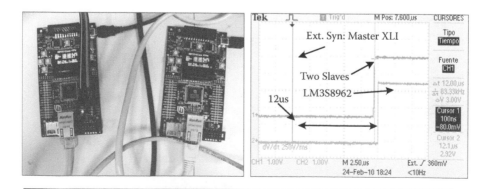

Figure 9.7   Stability for two slave LM3S8962.

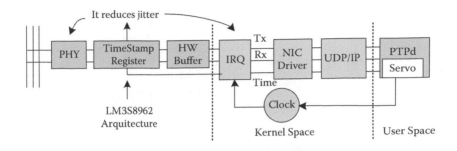

Figure 9.8   LM3S8962 architecture for IEEE 1588.

The software integrates the precision time protocol daemon (PTPd). The PTPd is a complete implementation of the IEEE 1588 specification for a standard nonboundary clock.

It was developed by two engineering students at Case Western Reserve University over approximately six months as part of an undergraduate senior project [14].

Figure 9.8 shows the message send and receive paths in a LM3S8962 microcontroller system running the PTPd [14]. For applications requiring very high-precision synchronization packets, the Ethernet controller provides a means of generating precision time stamps in support of the IEEE precision time protocol: IEEE 1588 [17]. This feature is enabled by setting the TSEN bit in the Ethernet MAC timer support (MATCS) register. A general purpose timer must be dedicated to the Ethernet controller for storing the receive time and the transmit time.

9.3.5 IEEE 1588 v2 2008

IEEE 1588-2008 has recently been developed to meet synchronization needs in new applications. New requirements have emerged as: higher accuracy, varied update rates, linear topology, and rapid reconfiguration after network changes and fault tolerance [24].

Furthermore, new applications in telecommunications and the power industry are an additional driving force of the discussion of the new version.

Several new features are proposed in IEEE 1588-2008 to meet with the objectives. For higher accuracy, a correction field of 64-bit integers is introduced, which consists of a 48-bit nanosecond part and a 16-bit subnanosecond part.

IEEE 1588 v2 devices that support the peer delay mechanism measure a peer delay for every communication link to rapidly reconfigure synchronization hierarchy. Moreover, the new version defines ANNOUNCE messages for the selection of the best master clock.

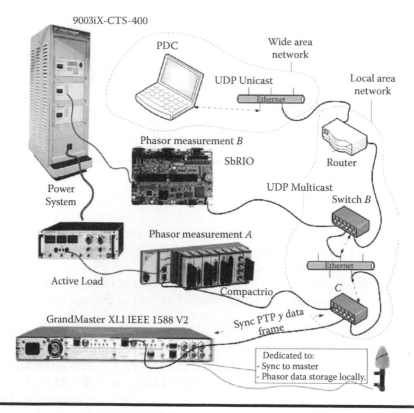

Figure 9.9 Experimental system for PTP v2.

9.3.6 *Experimental System for Test with PTP v2*

This system supports industrial I/O modules for power and energy measurement Figure 9.9. It is composed of a XLi grandmaster with GPS and two NI CompactRIO systems. For calibration of each of the experimental slaves, this system is based on the XLi's time interval/event time (TIET) feature, which can be used to measure PTP v2 synchronization across timing networks.

The two slave units work by EID system architecture: an architecture with NI RIO-9074 (400 MHz industrial real-time processor, 2 M gate, 8-slot FPGA chassis for custom I/O timing and processing) and another architecture with SbRIO-9631 (266 MHz industrial real-time processor, 1 M gate, with an assortment of analog, digital, and industrial I/O).

AC/DC power sources 9003iX-CTS with a high-performance power analyzer. It applies wave form precision for two phases simultaneously. The same slave transmits to the central device the frames as the norm synchrophasor standard [2]. The method must emulate the traffic in a substation.

National Instruments introduced a new clock available to NI-TimeSync. The IEEE Standard 1588 plug-in available with NI-TimeSync provides a clock reference that is synchronized with 1 ms resolution. You can configure multiple devices on a network to use the same IEEE 1588 reference clock, allowing multiple platforms to synchronize over a standard Ethernet network. You also can configure your device to use the software 1588 precision time protocol.

9.4 IEC 61850 Standard

IEC 61850 has been substantially developed for substations but is seen as a key standard for all field equipment operating under both real-time and non-real-time applications. IEC 61850 is better described as a communication system than a protocol. It includes parts for modeling of the components and the system, a description of data types and classes, abstract service definitions, specific mapping for system implementation, and conformance testing.

Edition V1 of 61850 only included communication within the substation, so connection to the control center was informational only and did not include 61850 services.

With the completion of Section 90-1 in 2009, methods for direct communication with 61850 outside of a single substation became a part of the standard and are fully described and supported.

9.4.1 Harmonization of IEEE C37.118 with IEC 61850 and Precision Time Synchronization

There are significant differences in scope and content of the two standards. IEEE C37.118 includes communication as well as measurement requirements, and is also intended to support applications such as protection. IEC 61850 is suitable for systemwide applications that require higher publishing rates. The approach including possible models for PMU data needs to be defined in IEC 61850 [19].

Common time synchronization will be a key for many smart grid applications. The IEEE 1588 standard will be a key element to achieve that synchronization, and the PC37.238 standard is developing for application of PTP to electric power [25].

It is possible to use a similar approach for the transmission of PMU and PDC data, but the capability needs to be formally defined in IEC 61850. PC37.239 defines a common format for event data exchange (COMFEDE) for power systems.

9.4.2 PC37.239 for Common Format for Event Data Exchange

This standard defines a common format for the data files needed for the exchange of various types of power network events in order to facilitate event data integration and analysis from multiple data sources and from different vendor devices. The flexibility provided by digital devices in recording network fault event data in the electric utility industry has generated the need for a standard format for the exchange of data. These data are being used with various devices to enhance and automate the analysis, testing, evaluation, and simulation of power systems and related protection schemes during fault and disturbance conditions. Since each source of data may use a different proprietary format, a common data format is necessary to facilitate the exchange of such data between applications. This will facilitate the use of proprietary data in diverse applications and allow users of one proprietary system to use digital data from other systems [26].

9.4.3 PC37.238 for Application of PTP to Electric Power

This standard specifies a common profile for use of the IEEE 1588-2008 precision time protocol (PTP) in power system protection, control, automation, and data communication applications utilizing Ethernet communications architecture [25].

The profile specifies a well-defined subset of IEEE 1588-2008 mechanisms and settings aimed at enabling device interoperability, robust response to network failures, and deterministic control of delivered time quality. It specifies the preferred physical layer (Ethernet), a higher-level protocol used for PTP message exchange, and the PTP protocol configuration parameters. Special attention is given to ensuring consistent and reliable time distribution within substations, between substations, and across wide geographic areas.

9.4.4 Experimental System for Test with PTP and IEC 61850

The experimental system shown in Figure 9.10 is a complete system ready to test measurement and real-time communications: the system is composed of a PCI1588 master (GPS), two LM3S8962 slaves, two PMUs, and two IPC@CHIP-based systems.

For calibration of each of the experimental slaves this system, based in XLi's time interval/event time (TIET) feature, can be used to measure PTP v2 synchronization across timing networks.

For calibration of each PMU this system is based in the AC/DC power source 9003iX-CTS with a high-performance power analyzer.

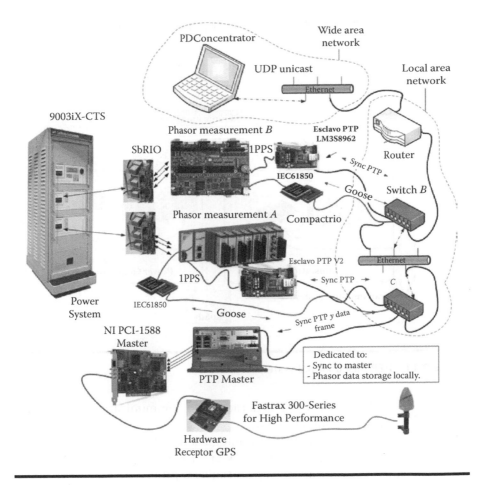

Figure 9.10 Experimental system for PTP v2.

For calibration of two IPC@CHIPs, the system is based in IEDScout software.

9.5 PTP-Based V2 Global System for Synchronized Event in Smart Grid

For event synchronizing in smart grid we use a PTP-based V2 global system that can provide a secure communication channel with a delay that does not compromise the correct operation of the global system.

This would imply the advantage of reusing the infrastructure of existing telecommunications networks to transmit synchronism information between PMUs. Figure 9.11 shows an example of generic application of PMUs

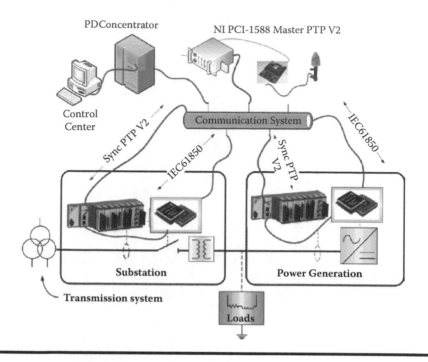

Figure 9.11 PTP-based global system.

in smart grids. IEC 61850-7-420 is dedicated to communications systems for distributed energy resources (DER).

The standardization enables the integration of the equipment and systems for controlling the electric power process into complete system solution, which is necessary to support utilities processes. Ensure the interoperability of equipment and systems by providing compatibility between interfaces, protocols, and data models. With IEC 61850's standardization of data acquisition and description methods, integration efforts are reduced [11].

The data concentration function also requires supporting a wide range of communications protocols. And they should support the newer standard protocols for both IEDs and SCADA masters. Standard protocols such as DNP3™, IEC 60870-5, and IEC 61850 (including GOOSE) may be needed now or in the future. When applicable, both serial and LAN formats should be specified. User-friendly features such as configuration templates for all protocols can reduce the configuration time considerably [12].

In addition, the network time protocol (NTP), simple network time protocol (SNTP), and precision time protocol (PTP) may be required to allow time synchronization over the network.

We also study the possibility of adding functionality to transmitting GOOSE messages on an Ethernet network, and the integration of the PTP

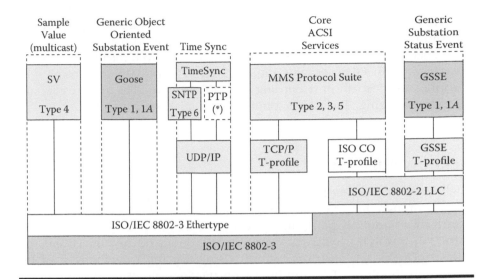

Figure 9.12   Protocol mapping profile.

protocol for synchronizing tasks (Figure 9.12) as proposed in the paper [11]. This scheme represents the IEC 61850 protocol mapping profile.

The OMICRON IEDScout software, which is available at [13], was used to detect and subscribe GOOSE messages on the network. Several GOOSE messages that were transmitted on the network were detected by the IED-Scout software.

For this task we use a development system for building applications based on the embedded web controllers IPC@CHIP SC123 and IPC@CHIP SC143. It runs with the @CHIP-RTOS operating system, which includes features like a real-time kernel, a full TCP/IP stack with a UDP/TCP socket interface, IEC 61850 integrated, and time synchronization via Simple Network Time Protocol (SNTP). Alternatively, we can use the IEC 61850/61400-25 dynamic link library (DLL) to communicate from a computer.

9.6 Conclusions

We have developed different methods for testing the synchronization with the IEEE 1588 standard. We tested the PTP synchronization in two PMUs and have proposed two methods: with the slave integrated into an embedded system, CompactRIO, and an external slave with an ARM microcontroller. In the second case, we have reached a synchronization with an error of 12 μs.

The use of a PTP-based global system for synchronizing smart grids provides a secure communication channel with a delay to within 12 μs. We use a master PCI-1588 and its stability around 100 ns.

The FPGA technology controls all task timing, capture, and preprocessing and provides a deterministic method for phasor estimation.

We study the possibility of adding functionality to transmitting GOOSE messages on an Ethernet network and the IEC 61850-90-5 to transmit synchrophasor information according to IEEE C37.118. The first results are quite encouraging. Today we continue with more specific tests.

Acknowledgments

This research was supported partially by the company Telvent Energy, Spain, through the project Malaga SmartCity under contract 12009028. SmartCity's budget is partly financed by the European Regional Development Fund (ERDF), with backing from the Junta de Andalucía and the Ministry of Science and Innovation's Center for the Development of Industrial Technology. The authors thank the Spanish Ministry of Industry, Tourism, and Trade for funding Project TSI-020100-2010-484, which partially supports this work. Our unforgettable thanks to the Spanish Ministry of Science and Innovation for funding the research project TEC2010-19242-C03-02.

References

1. Carta, A., Locci, N., Muscas, C., Sulis, S., A flexible GPS-based system for synchronized phasor measurement in electric distribution networks, IEEE Transactions on Instrumentation and Measurement, 57(11) 2008.
2. Li, H., Li, W., A new method of power system state estimation based on wide-area measurement system, Industrial Electronics and Applications 2009 (ICIEA 2009), pp. 2065–2069.
3. Du, X.W., Liu, D.C., Li, Y., Mixed digital/analog testing system for relay protection of power system, Industrial Electronics and Applications 2007 (ICIEA 2007), 2nd IEEE, pp. 407–410.
4. IEEE 37.118-2005, IEEE standard for synchrophasors for power system.
5. Lixia, M., Muscas, C., Sulis, S., Application of IEEE 1588 to the measurement of synchrophasors in electric power systems, ISPCS 2009, International IEEE Symposium on Precision Clock Synchronization for Measurement, Control and Communication Brescia, Italy, October 12–16, 2009.
6. IEEE 1344-1995 (R2001), IEEE standard for synchrophasors for power systems.
7. IRIG 2004, IRIG serial time code formats, telecommunications and timing group, Range Commanders Council, U.S. Army White Sands Missile.
8. Begovic, M., Novosel, D., Djokic, B., Issues related to the implementation of synchrophasor measurements, in 41st Annual Hawaii International Conference on System Sciences, 2008, pp. 164.

9. Jiang, Z., Computational intelligence techniques for a smart electric grid of the future, Lecture Notes in Computer Science, 5551, 1191–1201, 2009.

10. Martin, K.E., Hamai, D., Adamiak, M.G., Anderson, S., Begovic, M., Benmouyal, G., et al., Exploring the IEEE standard C37.118-2005 synchrophasors for power systems, IEEE Transactions on Power Delivery, 23(4), 1805–1811, 2008.

11. Liu, Y., Zivanovic, R., and Al-Sarawi, S., Marinescu, C., Cochran, R., A synchronized event logger for substation topology processing, Power Engineering Conference, 2009 (AUPEC 2009), pp. 1–6.

12. Gunther, E.W., Cybectec substation gateways, SmartGridnew.com, November 5, 2008.

13. IEDScout—Software tool for engineers work with IEC 61850, version 2.0, OMICRON electronics GmbH, 2009.

14. Precision time protocol daemon (PTPd), http://ptpd.sourceforge.net/.

15. Timing and synchronization systems, Planning for synchronization, http://zone.ni.com.

16. Yang, G.H, Wen, B.Y., A device for power quality monitoring based on ARM and DSP, Industrial Electronics and Applications, 2006 1ST IEEE Conference, pp. 1–5.

17. IEEE 1588-2002, IEEE Standard for a precision clock synchronization protocol for networked measurement and control systems. Institute for Electrical and Electronics Engineers, New York, 2002.

18. NIST Special Publication 1108, NIST framework and roadmap for smart grid interoperability standards, Release 1.0, Janury 2010.

19. PAP13 61850 C37118, Harmonize and synchronization, Time synchronization, IEC 61850 objects/IEEE C37.118 harmonization (6.1.2, 6.2.2), 2009.

20. Martin, K. E., Synchrophasors in the IEEE C37.118 and IEC 61850, 5th International Conference on Critical Infrastructure (CRIS), September 2010, 20–22.

21. Pallarés-López, V., Moreno-Muñoz, A., Polonio Torrellas, M., Moreno García, I., Gonzalez de la Rosa, J.J., Experimental IEEE1588-BASED system for synchronized phasor measurement in electric substation, ICIEA 2010, pp. 942–947.

22. NERC, Security guideline for the electricity sector: Time stamping of operational data logs, Approved by Effective Date: TBD Critical Infrastructure Protection Committee, version 0.995.

23. National Instruments, Understanding timing and synchronization, http://zone.ni.com.

24. Han, J., Jeong, D.K., A practical implementation of IEEE 1588-2008 transparent clock for distributed measurement and control systems, IEEE transactions on Instrumentation and Measurement, 59(2), 2010.

25. PC37.238 ^{TM}/D5.6 2 draft standard profile for use of IEEE 3 std. 1588 precision time protocol in 4 power system applications, February 2011.

26. IEEE PC37.239 ^{TM}/D05 2 draft standard for common format for 3 event data exchange (COMFEDE) for 4 power systems, May 2010.

Chapter 10

Quality of Service in Networking for Smart Grid

Xiaojing Yuan, Wei Sun, Dong Han, Jianping Wang, Chongwei Zhang, and Xiaohui Yuan

Contents

To realize the grand vision of smart grid, the networking of various power electronics devices and management systems in a smart grid plays a critical role. The data and information are expected to be exchanged freely among these components in a reliable and timely manner. Given its importance in society development, the requirements of quality of service (QoS) for such networking is much higher than those of existing wireline and wireless networks among computers and mobile devices. Existing networking protocols lack the mechanism to provide such QoS. In this chapter, we first review the networking QoS need of various stages of the smart grid, describe the existing QoS support of popular wireless networking protocols and their limitations, and describe a new mechanism that takes QoS requirements into the network design phase to provide the upper and lower bounds for end-to-end delay and throughput even though the process is stochastic. Using IEEE 802.15.4, we demonstrated how to implement such a mechanism in an existing wireless networking protocol and conducted experiments to show the effectiveness of such enhancement.

10.1 Overview of Smart Grid Architecture and Its Networking QoS Needs

In the United States and around the world, modernization of the electric power grid is a central part of the global efforts to increase energy efficiency, integrate renewable energy sources, reduce greenhouse gas emissions, and build a sustainable economy that ensures prosperity for current and future generations. The smart grid vision of the 21st century hinges on successfully adopting, integrating, and advancing existing communication and computing technologies within the power delivery infrastructure [1,2]. The core technological challenge is the development of integrated information and communication/networking technologies for monitoring and control of the power electronic devices. The enabled real-time bidirectional information flow will ensure quick response to and restoration from any local scale power outage to prevent it aggregating to larger-scale blackouts like those that happened in North America and Europe [2,3].

Figure 10.1 shows the architecture of an integrated information and networking system (INS) for the smart grid. Four major parts in a power grid interact in this architecture, which include power generation, transmission, distribution, and consumer. They support situational awareness for detection, response, protection, and control functions in these power grid units. Different functions require different levels of quality of service (QoS) with respect to timeliness, reliability, availability, and survivability. For example, the QoS requirements for the monitoring function are much lower than those for real-time control, whereas the timeliness requirement to support detection function is comparable to that supporting control function.

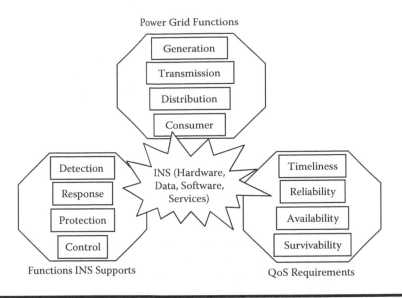

Figure 10.1   Smart grid architecture with information and networking systems.

The QoS for a power grid is twofold. First, it is desired to be able to deliver electricity at least 99.99999% of the time (7-nine level)—much more challenging than in any other system, especially networking systems, which typically require a 5-nine level of availability. Besides the disturbance and interruption to the power grid that can be detected and mitigated through the situational awareness of the INS, how to promptly recover from natural disaster also has significant impact, as revealed in the aftermath of Japan's recent earthquake. Second, when the system performs its functions and provides the desired services, a satisfactory quality of service needs to be maintained. Existing power grids' lack in this aspect makes the requirement of uninterruptible power supply (UPS) ubiquitous. The ultimate goal of integrating the information and networking system into the existing power grid to make it smart is to enhance the quality of service, in addition to making it available all the time (for example, to a 12-nine level).

With the situational awareness and automated control capability integrated into the power grid, we can greatly enhance the overall QoS of the power grid, such as sustaining the electricity quality when disturbance or interruptions occur and making the service more efficient, reliable, and secure. All these require a much higher quality of service standard for the two-way communication system than that used in the wired and wireless networks of computers and mobile devices. The high standards in the networking QoS are essential to ensure timely and reliable exchange of data and information among components in the power generation and delivery system in the normal operation environment, enable quick recovery from any hardware or software failures, recover from natural disaster faster, and deliver crucial data even when the resources are pushed to the limit.

In this chapter, we will focus on the mechanism developed to enhance the timeliness of the two-way communication by modifying existing wireless networking protocols and use IEEE 802.15.4 as an example to demonstrate the effectiveness of such an approach.

10.2 Existing Wireless Networking Protocols and Their QoS Mechanism

With the increase in the electricity demand and the complexity of the power transmission and distribution network, the amount of data and information needed to be communicated and used in the INS also increase significantly. Such demand could result in serious network congestion using the existing networking strategies. Networking technology to ensure real-time, reliable, efficient, and effective bidirectional information flow for the INS is crucial for the functions it supports. In particular, the power distribution system monitoring has not been feasible [2,4] because of the overhead involved in installing and maintaining a traditional wired communication (such as those through fiber, Ethernet, RS485/232, etc.), the variety of electronic power devices, and the complex terrain they locate. Power line carrier (PLC) technology had been explored to support the data transmission with high potential to become an alternative solution for two-way communication. However, the data transmitted through power line tends to be interfered with by the power load [5]. The operation of switching devices such as a high-voltage direct-current (HVDC) transmission line and a UPS also interferes with the PLC communication [6]. Many power distribution systems also utilized a public cellular wireless network (GPRS/CDMA/3G) [7,8] to monitor the power distribution devices. However, besides the high cost, it provides no QoS mechanism and has higher security vulnerability. For a secure INS of the power grid, a dedicated private wireless network is necessary. In this chapter, we examine the three popular wireless technologies (Zigbee based on IEEE 802.15.4, Wi-Fi based on the IEEE 802.11, and WiMax based on IEEE 802.16) and their QoS mechanisms before summarizing the enhancements they need to support the smart grid vision. Table 10.1 summarizes the properties of these wireless technologies, including the design target, maximum capacity, QoS support, and security mechanism.

10.2.1 Zigbee Based on IEEE 802.15.4

Zigbee is a low-power and low-cost wireless technology based on IEEE 802.15.4 that has been widely used in the wireless personal area network (WPAN). The standard only defines the physical layer (PHY) and

Table 10.1 Comparison of Wireless Technologies: Zigbee, Wi-Fi, and WiMax

Wireless Technology	Zigbee	Wi-Fi	WiMax
Design target	Low-Cost, low-power, low-rate WPAN	Indoor WLAN and small networks	Indoor or outdoor WAN, large network
Target application	No infrastructure, nearby devices	Indoor, high capacity, small network	Outdoor, high-capacity, large network
Range channel bandwidth	10~75 m 5 MHz	10~100 m 5 or 10 MHz	Up to 50 km 5, 10, 20, or 40 MHz
Maximum Capacity	250 kpbs	54 Mbps	45 or 80 Mbps
QoS mechanism	None	Limited to AP's local QoS, 4 types: voice, video, best effort, and background	Network-wide QoS, 5 types: UGS, ertPS, rtPS, nrtPS, and BE
Security measure	No security measure	Security at MAC with WEP encryption, etc.	MAC layer encryption and key exchange, via licensed spectrum

medium access control (MAC) portion of the data link layer (DLL) because it was designed for wireless communication within a limited range: 10 to 75 m. The standard also specifies the devices operate in the 2.4 GHz, 915 MHz (U.S.), and 868 MHz (Europe) industrial, scientific, and medical (ISM) bands with a transmit power capacity as 1 mW (0 dBm) and raw data rate up to 250 kbps. Zigbee supports beacon- and nonbeacon-enabled networks. In non-beacon-enabled networks, an unslotted carrier sense multiple access/collision avoidance (CSMA/CA) channel access mechanism is used. In beacon-enabled networks, time synchronization is critical for all the nodes in the network. Both mechanisms enable the nodes in the network to conserve energy when not actively transmitting and receiving data. Zigbee does not support any QoS or security mechanism.

10.2.2 Wi-Fi Based on IEEE 802.11

Wi-Fi is based on a set of standards IEEE 802.11 for implementing a wireless local area network (WLAN) in three frequency spectrums: the 2.4, 3.6, and

5 GHz bands. Currently, it is the most popular wireless technology to connect computers and mobile devices, including smart phones, MP3 players, printers, digital cameras, and laptops. Like 802.15.4, the set of standards in the 802.11 family also define the PHY and MAC layers. Based on the different radio specifications, the range covered by any device using Wi-Fi varies, 32 m indoors and 95 m outdoors, with raw data rate in the Mbps range, much higher than that of Zigbee. Variation of the Wi-Fi implementations may provide a wider coverage range and higher raw data rate. The transmit power capacity is limited to 20 dBm (100 mW). Like the nonbeacon mode of the Zigbee, the MAC layer of Wi-Fi is also based on CSMA/CA, limiting the scalability of the protocol. Wi-Fi started to introduce some security measures such as suppressing the access point's (AP) service set identifier (SSID) broadcasting and wired equivalent privacy (WEP) encryption in 802.11i. The QoS mechanism was introduced in IEEE 802.11e, including four types of data traffic: voice, video, best effort (BE), and background. The benefit is limited to the small network under one AP with QoS specifications.

10.2.3 WiMax Based on IEEE 802.16

WiMax (Worldwide Interoperability for Microwave Access) provides fixed and mobile Internet access based on the IEEE 802.16 family of standards, offering up to a 1 Gbps raw data rate, with the 802.16m update covering the range up to 50 km. While Zigbee and Wi-Fi support connectionless and contention-based access, WiMax runs a connection-oriented MAC. Thus, WiMax uses a QoS mechanism based on connections between the base station and the user device determined by the scheduling algorithm runs on the base station. This means the end user devices (i.e., the subscribers) cannot transmit data until a hit has been allocated a channel by the base station, and can only communicate to each other via base station. This allows WiMax to provide strong support of QoS for five different data traffic types: unsolicited grant service (UGS) for real-time data streams of fixed-size data packets, extended real-time pushing service (ertPS) for generating variable-sized data packet traffic periodically, real-time polling service (rtPS) for periodic variable-size data packets, non-real-time polling service (nrtPS) for delay-tolerant data streams with variable-size data packets, and best effort (BE), when no minimum service level is specified. WiMax also supports security features such as secure key exchange and encryption at the MAC layer. However, since it operates over a licensed spectrum, the connection is much more secure than that of Wi-Fi and Zigbee, which operate on ISM bands that are freely accessible to everybody. The big disadvantage of WiMax is its higher cost for setting up the base station and the end user device, as well as the energy consumptions for data transfer.

In summary, except for Zigbee, both Wi-Fi and WiMax support some QoS mechanism by predefining classes of traffic and the services associated

Table 10.2 Smart Grid Measurements Electrical Characteristics

Parameter Measured	Freq. Range (Hz)	Sampling Rate (Hz)	Resolution
Voltage	0.01 ⋯ 50–400	100–800	12~14 bits ADC
Current	0.01 ⋯ 50–400	100–800	
True power	0.01 ⋯ 50–400	100–800	14 bits ADC
Phase/frequency	5–15 +	10–30 +	0.01° phase
parameters derived			
Apparent power	0.01 ⋯ 50–400	100–800	14 bits ADC
Power factor	0.01 ⋯ 50–400	100–800	14 bits ADC
Accumulated energy	0.01 ⋯ 50–400	100–800	14 bits ADC

with them. Unfortunately, unlike the multimedia data exchanged through the Internet, most of the power grid data traffic (as summarized in Table 10.2) does not require the support for high bandwidth, but rather needs errorless delivery in a specified time period. In such cases, the overhead and the energy consumption associated with it will significantly increase the cost of communication within the INS. In the next section, we introduce a new QoS mechanism that provides more flexibility to the users so that they can define the QoS requirements of the networking based on specific application.

10.3 Flexible QoS Enhancement to Existing Wireless Networking Technologies

The QoS enhancement entails adding user-defined priority levels with respect to the application specification at the MAC portion of the data link layer of the chosen wireless networking protocol. Additional queues need to be added to the MAC, each queue serving traffic of the same priority level. Data in the high-priority queue not only have higher probability of channel access, but also can force service interruption of the lower-priority traffic. Without loss of generality, we consider two types of traffic, shown in Figure 10.2, going through the INS in the power grid: the *operational* data provides periodic measurements of the electricity quality and the device health condition—low priority; and the *emergency* data triggered by any detected failure or disturbance require immediate attention—high priority. Even though the operational data should be transmitted reliably at all times, emergency data should be able to preempt any operational data

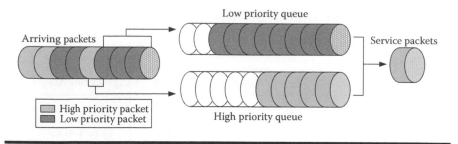

Low priority queue

Arriving packets

Service packets

□ High priority packet
■ Low priority packet

High priority queue

Figure 10.2 Queuing mechanism based on priority level of the data.

waiting for service, so that any failure can be addressed immediately to prevent a possible catastrophic event. Thus, high priority at the MAC layer indicates that emergency data will be given preferred service at all times over operational data.

Figure 10.2 shows two first-in-first-out (FIFO) buffer queues in the MAC layer for high- and low-priority data packets. The QoS-enhanced MAC follows two rules, one for service competition of data packets within a node, and the other for service competition of data packets among nodes in a network. Such a mechanism can easily be extended to traffic of more than two priority levels by adding queues at the MAC layer; higher-priority data can preempt any traffic at lower-priority levels following the same rules.

Rule 1: For the CSMA/CA mechanism in the node, any packets in the low-priority queue will not be serviced if the queue of the high-priority packets is not empty.

Rule 2: When nodes compete with each other for channel access in a network, the nodes with high-priority packets have a fixed shorter backoff period (NB), smaller backoff time (BE), and more frequent clear channel assessment (CCA) detection, while the nodes having low-priority packets use larger random backoff period, longer backoff times, and less frequent CCA detection.

These two rules are simple enough to be embedded into any networking stack without much additional expenses (i.e., overhead).

Since Zigbee has no QoS measure and is widely used in wireless sensor networks that have been applied in various power transmission and distribution monitoring systems, [2,9,11] we use IEEE 802.15.4 as an example to derive the delay model of high- and low-priority data based on queuing theory to capture the delay suffered by one application frame for successful one-hop transmission. System level performance measures such as goodput[1] and collision probability will also be studied.

---

[1] We define "goodput" as the ratio of achieved throughput and the traffic arrival rate.

The IEEE 802.15.4 MAC layer supports two medium access modes: the slotted mode (beacon-enabled mode) and unslotted mode (non-beacon-enabled mode). In the slotted mode, the MAC layer accesses the medium using a beacon-based superframe structure. In the *unslotted* mode, arbitration of medium access of distributed wireless nodes is achieved by the carrier sense multiple access/collision avoidance (CSMA/CA) scheme [11]. Studies [13,15] found that the unslotted CSMA/CA mode of IEEE 802.15.4 has better goodput and a higher successful transmission rate than the slotted mode. Unlike in the CSMA/CA of IEEE 802.11, in which the backoff counts down only when the wireless channel is idle, [11,12] in IEEE 802.15.4 CSMA/CA, the backoff period counts down all the time.

Table 10.3 summarizes the important symbols and notations we will use in the derivation of the delay model based on queuing theory. In the derivation, we assume a wireless sensor network that has N sensor nodes that monitor the electricity quality and health status of power electronic

Table 10.3 Symbols and Notations used in Analytical Model

Symbol	Description
n_0	Number of nodes with high priority packages
n_1	Number of nodes with only low priority packages
n_2	Number of nodes with no packages in MAC queue
T_s	Service time of a package
b_m	Average backoff time of the m-th stage
T_{tx}	Average transmission time of a packet
K	The maximum backoff stage of high priority packet
p_i	The probability of a channel is idle
$p_{i|i}$	The probability of a channel is idle in two consecutive periods
τ	The probability of CCA detection
p_0	The probability of a node with high priority package
p_1	The probability of a node with only low priority package
P_2	The probability of a node with no package
m_0	The buffer size for high priority queue
m_1	The buffer size for low priority queue
$Q_{n0,n1}$	The probability of having n0 nodes with high priority package and n1 nodes with only low priority package in a network

devices and report back to a coordinator using the IEEE 802.15.4 protocol. When abnormality is detected, it will be considered emergency data and pushed into the high-priority queue of the MAC layer. If no other high-priority data are being serviced at the time, it will interrupt the service of low-priority operational data packets, if there is any. That is, no operational data will be serviced until the high-priority queue is completed.

Assume the data packet arrival rates are λ_0 and λ_1 for high-priority and low-priority data packets, respectively. The nodes with high-priority packets $((\bullet)^0)$ are denoted as high-priority nodes,[2] and the nodes with *Only* low-priority packets $((\bullet)^1)$ are low priority nodes. In each node, the buffer size for high- (low-) priority queue is m_0 (m_1). In the following subsections, we will derive the models of MAC delay, channel service time, and other performance measures, such as goodput and packet delivery failure rate.

10.3.1 MAC Delay Based on Markov Chain Queuing Model

The MAC delay of a package is defined as the time taken from the packet entering the MAC layer queue to it being serviced. For a high-priority packet, MAC delay T_d^0 includes time it spent in the high-priority queue waiting for other high-priority data packets before being serviced (Equation (10.1)) and the CSMA/CA channel access competitive time, i.e., channel service time T_s^0.

$$T_d^0 = \sum_{k_0=1}^{m_0} \sum_{k_1=0}^{m_1} \left(\frac{p_{k_0,k_1}}{\sum_{k_0=1}^{m_0} \sum_{k_1=0}^{m_1} p_{k_0,k_1}} \cdot k_0 \cdot T_s^0 \right) \tag{10.1}$$

where P_k0, k_1 is the probability of having k_0 high-priority packets and k_1 low-priority packets in the queues; T_s^0 is the average service time of high priority. When there is no other packet in the queue, the current high-priority packet will be serviced immediately and the MAC delay is T_s^0.

For low-priority packets, the MAC delay includes not only the time it takes to service all high-priority packets and low-priority packets before current packets in the queue, but also the service time of any new incoming high-priority packets. Thus, the MAC delay of low-priority packet T_d^1 can be calculated as

$$T_d^1 = \sum_{k_1=1}^{m_1} \left[\sum_{k_0=0}^{m_0} \frac{p_{k_0,k_1}}{\sum_{k_0=0}^{m_0} \sum_{k_1=1}^{m_1} p_{k_0,k_1}} \cdot k_0 \cdot T_s^0 + \frac{p_{0,k_1}}{\sum_{k_0=0}^{m_0} \sum_{k_1=1}^{m_1} p_{k_0,k_1}} \cdot k_1 \cdot T_s^1 \right]$$
$$+ T_d^1 \cdot \lambda_0 \cdot T_s^0 \tag{10.2}$$

---

[2] Such nodes can also have low-priority data packets in the queue.

where, T_s^1 is the average service time of low-priority packets. λ_0 is the average arriving rate of high-priority packets of the queue. Substituting T_d^0 in (10.2) we get:

$$T_d^1 = \frac{1}{1 - \lambda_0 \cdot T_s^0} \sum_{k_1=1}^{m_1}$$

$$\times \left[\sum_{k_0=0}^{m_0} \frac{p_{k_0,k_1}}{\sum_{k_0=0}^{m_0} \sum_{k_1=1}^{m_1} p_{k_0,k_1}} \cdot k_0 \cdot T_s^0 + \frac{p_{0,k_1}}{\sum_{k_0=0}^{m_0} \sum_{k_1=1}^{m_1} p_{k_0,k_1}} \cdot k_1 \cdot T_s^1 \right]$$

$$(10.3)$$

To solve Equations (10.1) and (10.3), we use the Markov chain queuing model shown in Figure 10.3 to get the probability of having k_0 high-priority data and k_1 low-priority data P_{k_0,k_1}. Each state in the Markov chain queuing model can be represented by a vector $< k, \lambda, \mu, p >$, in which k represents the number of packets in the queue, λ is the arrival rate, μ is the service rate, and p denotes the probability of having k packets in the queue at the current state. As shown in Figure 10.3, all states in the model were classified as states that have (10.1) at least one high-priority packet in queue (p^0); (2) no high-priority packet in queue (p^1), i.e., all packets are low-priority data; and (10.3) no packets in queue (p^2), i.e., the queues are empty. They can

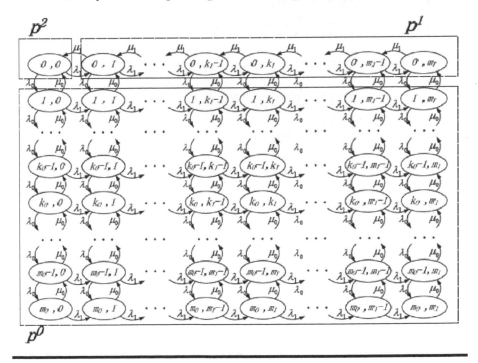

Figure 10.3 Markov chain queuing model for two MAC queues.

be determined based on the current states of the network (Equation (10.4)):

$$p^0 = \sum_{k_0=1}^{m_0} \sum_{k_1=0}^{m_1} p_{k_0,k_1}; \qquad p^1 = \sum_{k_1=0}^{m_1} p_{0,k_1}; \qquad p^2 = p_{0,0} \tag{10.4}$$

The probability of having n_0 nodes with high-priority packets and n_1 nodes with low-priority packets in queue, Q_{n_0,n_1}, can be calculated as

$$Q_{n_0,n_1} = \frac{N!}{n_0! \cdot n_1! \cdot n_2!} \cdot (p^0)^{n_0} \cdot (p^1)^{n_1} \cdot (p^2)^{n_2} \tag{10.5}$$

Figure 10.4 shows the derivation steps of solving the probability P_{k_0,k_1} ($k_0 = 1, 2, \ldots, m_0$ and $k_1 = 1, 2, \ldots, m_1$) under the conditions of statistical equilibrium [19].

$$
\begin{cases}
p_{0,0} = \dfrac{\mu_0}{\lambda_0 + \lambda_1} \cdot p_{1,0} + \dfrac{\mu_1}{\lambda_0 + \lambda_1} \cdot p_{0,1} \\[2ex]
p_{k_0,0} = \dfrac{\lambda_0}{\lambda_0 + \lambda_1 + \mu_0} \cdot p_{k_0-1,0} + \dfrac{\mu_0}{\lambda_0 + \lambda_1 + \mu_0} \cdot p_{k_0+1,0} \\[2ex]
p_{0,k_1} = \dfrac{\lambda_1}{\lambda_0 + \lambda_1 + \mu_1} \cdot p_{0,k_1-1} + \dfrac{\mu_1}{\lambda_0 + \lambda_1 + \mu_1} \cdot p_{0,k_1+1} \\[2ex]
\qquad + \dfrac{\mu_0}{\lambda_0 + \lambda_1 + \mu_1} \cdot p_{1,k_1} \\[2ex]
p_{k_0,k_1} = \dfrac{\lambda_1}{\lambda_0 + \lambda_1 + \mu_0} \cdot p_{k_0,k_1-1} + \dfrac{\lambda_0}{\lambda_0 + \lambda_1 + \mu_0} \cdot p_{k_0-1,k_1} \\[2ex]
\qquad + \dfrac{\mu_0}{\lambda_0 + \lambda_1 + \mu_0} \cdot p_{k_0+1,k_1} \\[2ex]
p_{m_0,0} = \dfrac{\lambda_0}{\lambda_1 + \mu_0} \cdot p_{m_0-1,0} \\[2ex]
p_{m_0,k_1} = \dfrac{\lambda_0}{\lambda_1 + \mu_0} \cdot p_{m_0-1,k_1} + \dfrac{\lambda_1}{\lambda_1 + \mu_0} \cdot p_{m_0,k_1-1} \\[2ex]
p_{0,m_0} = \dfrac{\lambda_1}{\lambda_0 + \mu_1} \cdot p_{0,m_1-1} + \dfrac{\mu_0}{\lambda_0 + \mu_1} \cdot p_{1,m_1} \\[2ex]
p_{k_0,m_1} = \dfrac{\lambda_0}{\lambda_0 + \mu_0} \cdot p_{k_0-1,m_1} + \dfrac{\mu_1}{\lambda_0 + \mu_0} \cdot p_{k_0,m_1-1} \\[2ex]
\qquad + \dfrac{\mu_0}{\lambda_0 + \mu_0} \cdot p_{k_0+1,m_1} \\[2ex]
p_{m_0,m_1} = \dfrac{\lambda_0}{\mu_0} \cdot p_{m_0-1,m_1} + \dfrac{\lambda_1}{\mu_0} \cdot p_{m_0,m_1-1}
\end{cases}
$$

Figure 10.4 Derivation step of solving the probability p_{k_0,k_1}.

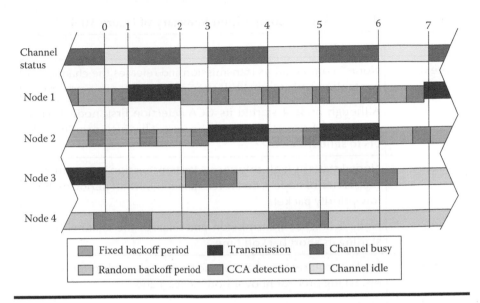

Figure 10.5 Channel access competition time diagram of CSMA/CA mechanism.

10.3.2 *Channel Service Time Model*

The time diagram in Figure 10.5 shows the mechanism of the CSMA/CA channel access competition. It shows the timeframes of four nodes contending the channel access using the CSMA/CA mechanism. Nodes 1 and 2 have high-priority packets to transmit, while nodes 3 and 4 have only low-priority packets. At the beginning, packets from node 3 are being serviced. Later, nodes 1 and 2 receive their high-priority packets and initiate the CCA detection to service these packets. Since the channel is busy, a backoff period is initiated. Table 10.4 summarizes the time periods the four nodes went through.

In summary, as illustrated in Figure 10.5, it is possible that when multiple nodes are servicing high-priority data packets, some of them still will be forced to back off and wait. For example, node 1 has to wait from time 2 to time 7 because its backoff period finished before the channel is released, and node 2 was forced to back off at time 7.

To model the channel access competition of the CSMA/CA mechanism, we adopted the time unit "symbol" (1 *symbol* = 16 *ns*) as accepted in the networking community. The backoff period, turnaround time, and other time used in the MAC layer are specified in the unit "symbol" instead of "second." Consider two continuous time units in the channel; the probability of the channel being idle (P_i) can be calculated as $P_i = P_{i|i} * P_i + P_{i|b} * (1 - P_i)$, in which $P_{i|b}$ is the conditional probability that the channel is idle in the second time unit given it is busy in the first time unit, and $P_{i|i}$ is the

Table 10.4   Channel Access Competition Summary of Figure 10.5

Time	Description
Time 0	Node 3 completed its transmission and releases the channel for contention and automatically begins backoff for its next turn.
Time 0 to 1	Although node 4 started its CCA detection first, node 1 has a shorter CCA detection interval because it has high-priority packets to send (rule 2).
Time 1	Node 1 starts its high-priority packet transmission by interrupting the service of node 4, and forces another backoff period of these low-priority packets.
Time 2	Node 1 finishes transmission and releases the channel before starting its short backoff time.
Time 3	Node 2 acquires the channel and transmits its high-priority packets, taking precedent over both nodes 3 and 4.
Time 4	Node 4 finishes its backoff period and starts CCA detection right after node 2 finishes transmission and releases the channel. At this time, node 1 still detects a busy channel because it performs CCA detection before node 2 releases the channel.
Time 5	Node 2 sends its high-priority packets again since the CCA detection period and the backoff period of nodes 3 and 4 (both servicing low-priority packets) have been increased to ensure the channel is open for nodes servicing high-priority data. The chance of high-priority packets being dropped is rare.
Time 6	All the nodes either started their backoff period or CCA or have no data need to be transmitted.
Time 7	Node 1 starts to transmit its data, while node 2 is going through the forced backoff period since its CCA detection finishes later than that of node 1

conditional probability that the channel is idle in both time units. Given $P_{i|b} = 1/Avg(T_{tx})$ [15] and $Avg(T_{tx})$ denotes the average transmission time of a packet, we can calculate the probability of the high- and low-priority nodes detecting an idle channel for the packets in queue as

$$P_i^0 = 1/\left[1 + \overline{T}_{tx}\left(1 - P_{i|i}^0\right)\right]; \qquad P_i^1 = 1/\left[1 + \overline{T}_{tx}\left(1 - P_{i|i}^1\right)\right] \qquad (10.6)$$

If a node performs CCA in the first time unit when the channel is idle, it will send the data during the second time unit, making the channel busy. Only when no node performs the CCA in the first time unit is the probability of the channel being idle in both time units, $P_{i|i}$, nonzero. For high-priority nodes, the probability of an idle channel in two successive time units, $P_{i|i}^0$, is when no CCA is performed by any nodes in the first time unit (Equation (10.7)).

$$P_{i|i}^0 = \sum_{n0=1}^{N} \sum_{n1=0}^{N-n0} \left[\frac{Q_{n0,n1}}{Q_0} \cdot (1 - \tau^0)^{n0-1} \cdot (1 - \tau^1)^{n1} \right] \tag{10.7}$$

in which the probability of having at lease one high-priority node is $Q_0 = \sum_{n0=1}^{N} \sum_{n1=0}^{N-n0} Q_{n0,n1}$. The fraction in Equation (10.7) denotes the probability of having n_0 high-priority nodes and n_1 low-priority nodes in the network ($Q_{n0,n1}$) when there is at least one-high priority node (Q_0).

Similarly, the probability of an idle channel in two successive time units for low-priority nodes, $P_{i|i}^1$, can be calculated as

$$P_{i|i}^1 = \sum_{n0=1}^{N} \sum_{n1=0}^{N-n0} \left[\frac{Q_{n0,n1}}{Q_1} \cdot (1 - \tau^0)^{n0} \cdot (1 - \tau^1)^{n1-1} \right] \tag{10.8}$$

in which $Q_1 = \sum_{n_0=0}^{N} \sum_{n_1=1}^{N-n_0} Q_{n_0,n_1}$.

The probability of a node performing CCA detection per symbol for high-priority data (τ^0) or low-priority data (τ^1), in Equation (10.7) and (10.8), can be calculated using Equation (10.9) and (10.10), respectively:

$$\tau^0 = \sum_{n=0}^{K^0-1} \frac{(1 - P_i^0)^n \cdot P_i^0 \cdot (n+1)}{\sum_{m=0}^{n} (b_m^0 + 1) + T_{tx}^0} + \frac{(1 - P_i^0)^{K^0} \cdot K^0}{\sum_{m=0}^{K^0-1} (b_m^0 + 1)} \tag{10.9}$$

$$\tau^1 = \sum_{n=0}^{K^1-1} \frac{(1 - P_i^1)^n \cdot P_i^1 \cdot (n+1)}{\sum_{m=0}^{n} (b_m^1 + 1) + T_{tx}^1} + \frac{(1 - P_i^1)^{K^1} \cdot K^1}{\sum_{m=0}^{K^1-1} (b_m^1 + 1)} \tag{10.10}$$

The first term in both equations represents the average CCA detection probability when the packet is transmitted successfully: CCA detection time divided by the sum of successful transmission time of the packet (T_{tx}^0), the time for n backoff stages (in which b_m^0 is the average backoff time of the m th backoff stage), and one unit time for CCA detection.

A packet is serviced either when the packet is transmitted successfully or when it is dropped due to CCA detection failure and the maximum backoff time has been reached. The channel service time of high- and low-priority packets can then be computed as follows:

$$T_s^0 = \sum_{n=0}^{K^0-1} (1 - P_i^0)^n \cdot P_i^0 \cdot \left[\sum_{m=0}^{n} (b_m^0 + 1) + T_{tx}^0 \right] + (1 - P_i^0)^{K^0} \cdot \sum_{m=0}^{K^0-1} (b_m^0 + 1) \tag{10.11}$$

$$T_s^1 = \sum_{n=0}^{K^1-1} \left(1 - P_i^1\right)^n \cdot P_i^1 \cdot \left[\sum_{m=0}^{n} \left(b_m^1 + 1\right) + T_{tx}^1\right] + \left(1 - P_i^1\right)^{K^1} \cdot \sum_{m=0}^{K^1-1} \left(b_m^1 + 1\right)$$

(10.12)

The first term is the sum of the time it takes to transmit a packet within n backoff stages: the probability that the packet is transmitted successfully multiplies the sum of the transmission time of the packet. The second term is the time spent when the packets failed to transmit. Note that the unit of time is "symbol."

10.3.3 MAC Delay, Goodput, and Packet Delivery Failure Rate

In this section, we summarize the derivation from previous two subsections and derive the equations for performance measures such as the MAC delay, goodput, and collision rate.

We consider the MAC delay as the time taken from when the packet enters the MAC layer queue to it being serviced, i.e., either transmitted successfully or dropped. Given the channel service times T_s^0 and T_s^1 as well as the Markov chain model of the MAC queues, we can calculate the MAC delay for high- and low-priority packets using Equations (10.1) and (10.3), respectively.

The goodput of periodic operational data (G_1) through the network indicates whether the data of power quality and health status of the electronic power devices can be reliably transmitted during the normal operation. Likewise, the goodput of bursty emergency data (G_0) ensures that any failures of the power grid can be detected and get responded to in time to prevent escalated damage. In addition, the ratio between G_1 and G_0 is also an important factor in the networking characteristic of the smart grid. Given network parameters, we can calculate the goodput values for high- and low-priority data as

$$G_0 = \frac{\sum_{n_0=0}^{N} \sum_{n_1=0}^{N-n_0} Q_{n_0, n_1} \cdot \left(n_0 \cdot \frac{\left[1-\left(1-P_i^0\right)^{K_0}\right] \cdot L_0}{T_s^0}\right)}{N \cdot \lambda_0}$$

(10.13)

$$G_1 = \frac{\sum_{n_0=0}^{N} \sum_{n_1=0}^{N-n_0} Q_{n_0, n_1} \cdot \left(n_1 \cdot \frac{\left[1-\left(1-p_i^1\right)^{K_1}\right] \cdot L_1}{T_s^1}\right)}{N \cdot \lambda_1}$$

(10.14)

in which L_0 and L_1 denote the average packet sizes, and K_0 and K_1 denote the maximum backoff periods of high- or low-priority data, respectively.

In the smart grid, the packet transmission failure rate indicates the reliability of the periodic operational data communication. Ideally, all operational data should be transmitted even though they may be delayed when bursty emergency traffic occurs. Packet dropping is caused by the inability of the packet to use the wireless channel in a timely manner or collision at the coordinator. Given the network parameters, we can calculate the packet transmission failure for high- (F_0) and low- (F_1) priority data as

$$F_0 = \left(1 - P_i^0\right)^{K^0} \tag{10.15}$$

$$F_1 = \left(1 - P_i^1\right)^{K^1} \tag{10.16}$$

10.4 Experiment Design and Results

In this section, we present the experiment design and results for the enhanced IEEE 802.15.4 to validate and demonstrate the effectiveness of the quality of service enhancement mechanism.

The performance of the QoS-enhanced MAC with respect to network delay, goodput, and collision rate is being examined with a wireless network consisting of 10 sensor nodes and 1 coordinating node that simulate the monitoring system of a power distribution substation. The maximum data rate at 2.4 GHz PHY of IEEE 802.15.4 is 250 kbps. The packet size is 50 bytes and the transmission time (including acknowledgment and interframe space (IFS)) is 157 symbols. The default values of *UnitBackoffPeriod* (20 symbols) and *TurnaroundTime*[3] (12 symbols) are used in the simulation. To avoid interruption of the acknowledgment (ACK) frame, we limited the duration for CCA detection to one *UnitBackoffPeriod* (i.e., 20 symbols). The packet arriving rates for all nodes are the same. The buffer sizes of MAC queues for both high- and low-priority data are the same at 300 bytes. The maximum number of backoff stage (BN) is 5. The backoff exponent (BE) ranges from 0~3 for high-priority data packets and 2~5 for low-priority data packets.

We designed two test scenarios to study the network behavior of both types of traffic for when there is (1) sufficient bandwidth and (2) saturated bandwidth—demonstrating the performance boundary for high-priority data using the QoS-enhanced Zigbee protocol. Since the high-priority data are typically bursty emergency data and low-priority operational data are continuous periodic monitoring traffic, we analyzed the network delay, goodput, and collision rate for when the bursty emergency data arriving rate (λ_0) is (a) 8 kbps and (b) 16 kbps, while varying the low-priority data arriving rate (λ_1) from 0.5s to 23 kbps, with 1 kbps increment on each

---

[3] The interval between the transmission (TX) frame and acknowledgment (ACK) frame.

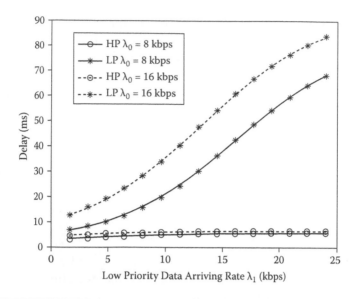

Figure 10.6 Network behavior comparison between high- and low-priority data: Network delay.

node. As the result, the maximum traffic through the network for test scenario (a) ranges from 8 kbps to 310 kbps, and 16 kbps to 390 kbps for test scenario (b). In both scenarios, all the nodes try to send generated traffic simultaneously, thus simulating the worst-case scenario in a networking perspective. When the data traffic in the network exceeds the maximum bandwidth supported by IEEE 802.15.4 (250 kpbs), the quality of service affects all network performance measures as shown in Figures 10.6 to 10.8.

Figure 10.6 shows a comparison of network behavior between the high- and low-priority data with respect to the network delay. We observe that (1) the delay of high-priority data remains below 5 ms and invariant when the traffic in the network goes well beyond the capacity of the protocol support; (2) the delay of low-priority data increased significantly with the increasing of the traffic in the network, from around 10 to around 80 ms, because of the increased backoff stages low-priority data need to go through when preempted by the high-priority data; and (Figure 10.3) the increase in the network delay with the high-priority traffic arriving rate increasing from 8 to 16 kbps is proportional for the low-priority data, while negligible for the high-priority packets.

The goodput for both scenarios is shown in Figure 10.7. It is clear that the goodput of low-priority data dropped sharply (from around 90% to 30%), while the goodput for high-priority data degraded in a relative slower pace (from nearly 100% to 65%). When the traffic is very heavy, less than 30% low-priority data will be delivered successfully, while more

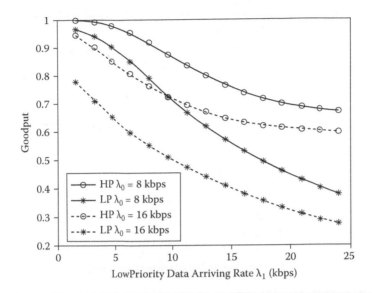

Figure 10.7 **Network behavior comparison between high- and low-priority data.**

than 60% emergency data transmitted successfully. It demonstrates that the QoS-enhanced IEEE 802.15.4 MAC layer ensures quality of service for high-priority data by sacrificing some low-priority packets. Note that we expect such behavior in a failure so that timely responses can be made to prevent exacerbation of the situation.

Figure 10.8 shows the collision rate of both high- and low-priority data after the maximum backoff stage is reached and the packet is dropped. When the traffic is light, the collision rate of both priority packets is small (less than 5%). With the increase in traffic, the collision rate for both priority data increases because the channel is busy. However, the collision rate for the low-priority data increases much faster than that of the high-priority data, demonstrating the preemptive QoS enhancement to the CSMA/CA for the high-priority data over low-priority data.

Our simulation experiment results using the QualNet [19,20] network simulator confirm the trend of all three measures: network delay, goodput, and collision rate. As shown in Figure 10.9, even though the average network delay of the low-priority data (7 ms) is approximately two times that of the high-priority data (3.8 ms), the ratio between their variance is about 10 (low priority:high priority). The simulation results demonstrated the effectiveness of the QoS enhancement mechanism. For high-priority data, the quality of service will not degrade as much as that of the low-priority data, and have a deterministic boundary for all three QoS measures.

In summary, the experiment results are based on an analytical model and simulation cross-validated with each other, and demonstrate the

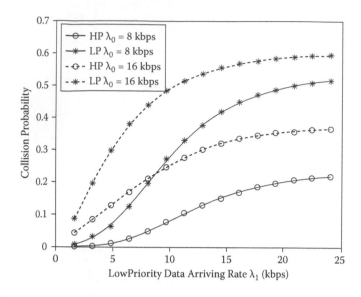

Figure 10.8 Network behavior comparison between high- and low-priority data.

effectiveness of the QoS enhancement to the MAC portion of the data link layer for existing wireless technology. With the stochastic networking behavior, there exists an upper boundary for all three performance measures. In addition, as the amount of traffic in the network increases, the performance disparity between high-and low-priority data also increases.

10.5 Conclusions and Outlook

In this chapter, we identified the networking quality of service requirements, namely, the timeliness, reliability, availability, and survivability, for the INS to support functions such as detection, response, protection, and control, which enable the smart grid vision. We reviewed the QoS support of existing wireless technologies, such as Zigbee, Wi-Fi, and WiMax. These wireless technologies have been researched to support various stages of the power grid monitoring and control, including power generation, transmission, distribution, substation monitoring, and consumer end. The timeliness requirement of the network quality of service is the most fundamental, since the reliability, availability, and survivability of a network all depend on whether the packets can be delivered in time to the right nodes. This chapter focuses on presenting an enhancement mechanism to improve the timeliness of the QoS by adding multiple queues at the MAC portion of the data link layer (DDL) of the network stack, and letting the user design

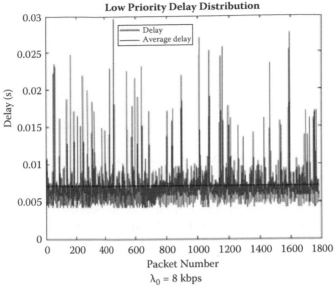

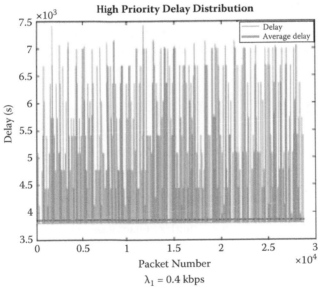

Figure 10.9   Network delay distribution from QualNet simulation.

the priority levels based on specific application. Without loss of generality, theoretical equations of the MAC delay and network delay, goodput, and packet delivery ratio are derived based on two priority levels added to IEEE 802.15.4 contention-based CSMA/CA. Experiment results demonstrated the effectiveness of the mechanism in terms of network delay, goodput, and collision rate for two priority levels, as well as the feasibility of adding such an enhancement mechanism to the network stack of existing wireless technology. In summary, our QoS enhancement mechanism provides a uniform and deterministic boundary for MAC delay of emergency data, irrespective of the routine traffic competition.

Experiment results also show that the low-priority level traffic suffers a lot with respect to the network delay, goodput, and collision rate. The decision of whether or not the low-priority level data should be suppressed this much is up to the system designer. Sometimes, the emergency data about the potential catastrophic events do need to suppress all other traffic on the network, while other times routine traffic should be given certain bandwidth.

Future work on QoS enhancement mechanisms with respect to reliability, availability, and survivability is needed in order to enable the smart grid vision.

Acknowledgments

This work was supported by the National Science Foundation, the University of Houston, and the Hefei University of Technology.

References

1. *NIST smart grid standards*, National Institute of Standards and Technology working draft proposed standard, rev. release 1.0 (2009).
2. V. C. Gungor, B. Lu, and G. P. Hancke, Opportunities and challenges of wireless sensor networks in smart grid—A case study of link quality assessments in power distribution systems, *IEEE Transactions on Industrial Electronics*, 1 (2010).
3. W. Sun, J. Wang, C. Zhang, and Z. Qian, Research on the wireless ad-hoc network for power distribution network communication in medium-small cities, in *International Conference on Sustainable Power Generation and Supply, 2009. SUPERGEN '09.* (2009).
4. J. Misurec, Interference in data communication over narrow-band PLC, *International Journal of Computer Science and Network Security*, 8 (2008).
5. S. Mahmod, N. F. Mailah, and M. Radzi, Effects of high power converter on power line carrier signal, in *4th National Conference on Telecommunication Technology*: 163–167 (2004).
6. P. K. Lee and L. L. Lai, A practical approach to wireless GPRS on-line power quality monitoring system, in *Power Engineering Society General Meeting* (2007).
7. A. J. Wilson, The use of GPRS technology for electricity network telecontrol, in *Computing & Control Engineering Journal* (2005).

8. R. A. Leon, V. Vittal, and G. Manimaran, Application of sensor network for secure electric energy infrastructure, in *IEEE Trans. on Power Delivery*, 22, 1021–1028 (2007).
9. M. M. Nordman and M. Lehtonen, A wireless sensor concept for managing electrical distribution networks, in *Power Systems Conference and Exposition, IEEE PES* (2004).
10. *Wireless medium access control and physical layer specifications for low-rate wireless personal area network*, IEEE 802.15.4, (2003).
11. Jung, C. Y., Hwang, H. Y., Sung, D. K., and Hwang, G. U., Enhanced Markov chain model and throughput analysis of the slotted CSMA/CA for IEEE 802.15.4, *IEEE Trans. on Vehicular Technology*, 58, 473–478 (2009).
12. F. Wang, D. Li, and Y. Zhao, Analysis and comparison of slotted and unslotted CSMA in IEEE 802.15.4, in *Wireless Communications, Networking and Mobile Computing, WiCom '09* (2009).
13. T. O. Kim, J. S. Park, H. J. Chong, and K. J. Kim, Performance analysis of IEEE 802.15.4 non-beacon mode with the unslotted CSMA/CA, in *Communications Letters, IEEE*, 238–240 (2008).
14. Y. K. Huang, A. C. Pang, and H. N. Hung, A comprehensive analysis of low-power operation for beacon-enabled IEEE802.15.4 wireless networks, *IEEE Transactions on Wireless Communications*, 5601–5611 (2009).
15. *Wireless LAN medium access control (MAC) and physical layer (PHY) specifications: MAC quality of service enhancements*, IEEE 802.11e (2005).
16. *Distributed medium access control (MAC) for wireless networks*, WiMedia Alliance, http://www.wimedia.org (2005).
17. *Physical and MAC layers for combined fixed and mobile operation in licensed bands*, IEEE 802.16e (2005).
18. B. I. Marks, State probabilities of M/M/1 priority queues, *Operat Res*, 21 (1973).
19. K. A. Shuaib, A performance evaluation study of WiMAX using qualNet, in *Proceedings of the World Congress on Engineering 2009*, London, Vol. 1, July 1–3, 2009.
20. QualNet, http://www.scalable-networks.com/products/qualnet/.

Chapter 11

Communication Technologies, Networks, and Strategies for Practical Smart Grid Deployments: From Substations to Meters

Alberto Sendin

Contents

The smart grid concept is still in the process of being materialized. However, deployments of smart grid networks and systems are being launched worldwide. These smart grid deployments are still very much focused in smart metering, but there are cases of utilities that are already taking advantage of the deployment of telecommunication means at different levels of the grid, to evolve their implementations toward enhanced monitoring capabilities and improved automation systems in the grid.

Smart grid needs and capabilities are different in the various segments of the grid. This situation is due to the fact that the main enablers of smart grids are information and communication technologies (ICT), with communications probably being the bottleneck in ICT for smart grids. Information technologies do have to evolve from the existing ones today, due to the massive amounts of information that will have to be managed, and Communications, the focus of this text, must provide services in locations not typically covered by public systems, and for which traditional utilities' systems arc not prepared in terms of quantities, throughput, and coverage.

The need to use a single telecommunications network, blending public and private telecommunications technologies, is evidenced for each of the segments of the electricity grid.

11.1 Introduction

No two smart grids are the same. This is a bold statement that could lead us to two conclusions: the first one is that an important effort must be made to make the smart grid environment one in which utilities can share, rather than differentiate; and the second one, tightly joined to the first, is that common characteristics must be found to make this sharing of common practices and standards feasible.

The smart grid concept is getting mature [1–4], although it can still be considered quite new. This immaturity is partially caused by the lack of stable references of companies "already there," and partially because the circumstances of utilities are so different that even though use cases may seem similar, implementations are going to differ fundamentally.

Making an effort to find the common aspects of all smart grid views in the sector, it is possible to find a reference to the modernization of infrastructure, and the great increase in the number of sensors and controls in the electricity system. These elements, which have always been found in the grid, have experienced an evolution coming from the times when they were first introduced in the network, and monitored and operated at a local level, progressively taken care from distant and centralized control centers, and that will ultimately be part of automated systems capable of making complex decisions based on the overall data coming from the whole system. Thus, even if new infrastructure (grid) and algorithms (intelligence) are fundamental for smart grids, the glue that is going to make it all possible is ICTs [5–10], and specifically the infrastructure side of this newcomer, telecommunications.

However, telecommunications is not new for utilities. In fact, many telecommunications carriers today base their networks on basic infrastructure provided by these companies (rights of way, ducts, poles, fiber optics, telecommunication transmission services, etc.), and not a few of them are even spin-offs from utilities. The history of utilities is full of telecommunications networks being created due to the lack of public networks, or inadequacy of those networks to comply with the requirements of a business that is fundamental by itself (imagine the need to use telecommunications for substation repair purposes, when these telecommunications services are provided by a third party that bases them on the electricity supplied by that same faulty substation; the classic "chicken and egg" situation).

Even if telecommunications networks are today a fundamental asset for utilities when dealing with operational business in hundreds of thousands of premises spread over electricity distribution areas, communication access cannot probably be obtained in an economically affordable way with the traditional utility communications standard approach. The situation is even clearer if we consider the range of different service situations, where coverage of telecommunications services will not be systematically and inexpensively ensured by any single technology, but through a situation-specific mix of private and public telecommunications solutions. In both cases, and for the combination of networks needed, the challenge will be to combine seamlessly an important number of different technologies.

This paper focuses on a practical approach to smart grid telecommunications, from the perspective of the ubiquitous telecommunications services needed in smart grids. These telecommunications services will be needed in millions of premises where utilities are present, and are a serious challenge for regular public telecommunications carriers: the nature of the services (real time, reliability, etc.), the varying capacity (bits per second), and the ubiquity of their presence challenge the design and dimensioning of many of the telecommunications networks in place today.

This chapter will not be an enumeration and description of technologies to be used (many could be identified, each at a different level, among them and not exhaustively, in public network domains, radio technologies and

wired technologies—xDSL, HFC, GPRS, UMTS, xDSL, HFC, etc.– and in private environments, technologies such as power line communications (PLC),
fiber optics, radiolinks, and radio coverage, and a range of mixed ones such
as WiMAX, PON, traditional analog and digital radio, and others), but an
identification of service scenarios according to the different segments of the
electricity system, with an outline of strategies to provide telecommunications services needed with a future-oriented vision, but without forgetting
the existing constraints.

11.2 Communications in Smart Grids

In an electricity system full of sensors and controls, the former are going to
give us lots of data, and the latter are going to give us the ability to react.
Imagine a system full of small islands, isolated and unable to exchange data
and forced to make decisions only based on their own shortsighted view
of the situation at the local level. Imagine the uncoordinated behavior and
its consequences.

Telecommunications must make smart and global perspective-based decisions a reality. Additionally, telecommunications will make remote or
automated control possible, in an environment where one or the other
approach is now probably a consequence of viability, evolution, or even
confidence in the telecommunication means available Table 11.1.

Smart grid-associated telecommunications services will be defined in
terms of the following characteristics:

- Depending on the distance involved, services can be of the local,
  regional, national, or international level.
- Depending on the bandwidth involved, services can be narrowband
  or broadband. This aspect will measure the bits per second we can
  deliver in the channel.
- Depending on the maximum delay the communication could admit,
  the maximum latency affordable for the service should be specified.
  As a very closely related parameter, the service will also be specified
  in terms of jitter, measuring the variation of the latency from one
  communication attempt to the next.
- Depending on the nature of the media or telecommunications technology used, the service will make use of a dedicated or shared capacity.
  Dedicated capacity will always be there to be used with the same
  characteristics (up to a certain limit), independently of the number
  of users present in the system. This will not be the case with shared
  capacity technologies.
- Depending also on the nature of the telecommunications infrastructure supporting the service, it will be delivered over wired or wireless

Table 11.1　Different Definitions for Smart Grids

Organization/ Author	Grid/Concept	Definition
Jeju Smart Grid Project (2009)	Jeju Island (Korea) test bed	A smart grid refers to a next-generation network that integrates information technology (smart) into the existing power grid (grid) to optimize energy efficiency through a two-way exchange of electricity information between suppliers and consumers in real time.
Climate Group (2008)	Smart grid	A smart grid is a set of software and hardware tools that enable generators to route power more efficiently, reducing the need for excess capacity and allowing two-way, real-time information exchange with their customers for real-time demand side management (DSM). It improves efficiency, energy monitoring, and data capture across the power generation and T&D network.
Adam and Wintersteller (2008)	Smart grid	A smart grid would employ digital technology to optimize energy usage, better incorporate intermittent "green" sources of energy, and involve customers through smart metering.
Miller (2008)	Smart grid	The smart grid will: • Enable active participation by consumers • Accommodate all generation and storage options • Enable new products, services, and markets • Provide power quality for the digital economy • Optimize asset utilization and operate efficiently • Anticipate and respond to system disturbances • Operate resiliently against attack and natural disaster

(*continued*)

Table 11.1   Different Definitions for Smart Grids (Continued)

Organization/ Author	Grid/Concept	Definition
European Technology Platform SmartGrids (2006)	Smart grid vision	Electricity networks that can intelligently integrate the behavior and actions of all users connected to them—generators, consumers, and those that do both—in order to efficiently deliver sustainable, economic, and secure electricity supplies.
EPRI (2005)	IntelliGrid	The IntelliGrid vision links electricity with communications and computer control to create a highly automated, responsive, and resilient power delivery system.
U.S. DOE (2003)	Grid 2030	Grid 2030 is a fully automated power delivery network that monitors and controls every customer and node, ensuring a two-way flow of electricity and information between the power plant and the appliance, and all points in between. Its distributed intelligence, coupled with broadband communications and automated control systems, enables real-time market transactions and seamless interfaces among people, buildings, industrial plants, generation facilities, and the electric network.

(radio frequency) networks. Both of them have different constraints in terms of the way the terminals connect to the system:

— Wired networks can be implemented based on metallic (copper cables, power lines, etc.) or dielectric cables (mainly fiber optic cables).

— Wireless networks will be located in different frequency bands, mainly either VHF/UHF or microwaves.

■ Depending on the nature of the carrier behind the service, the network can be either public or private, and in the case of public systems, the rest of nonutility users and the service characteristics will affect services provided for utilities.

Telecommunications services are not independent from the existing networks supporting them, and cannot go further than state of the art. This statement, although obvious, must be remembered when designing smart networks with a top-down approach. The issue here is that the smart grid from an ICT perspective will be based on applications that will support themselves on telecommunications, to reach sensors and controls in distant electricity system premises. If we design a system composed of applications and telecommunications with a top-down approach (i.e., considering applications alone), we are running a verifiable risk of imposing requirements on the telecommunications that are either beyond the state of the art or out of scope of feasible investments and expenditures, or simply not supported in today's telecommunications networks. With this possibility in mind, another risky situation could be to consider that the method to follow when designing ICT for the smart grid is the bottom-up approach, considering the limitations of today's networks, and the limited investment possibilities, and design or use of too pragmatic applications and telecommunications networks. This approach is also one that has its limitations, due to the fact that from the telecommunications perspective, it will not stimulate the creation and evolution of networks beyond their possibilities today, or alternatively, will just be supported by existing commercial networks. And more important, the investments today will have to be used for decades to come, and must consequently be future-proof.

The way to counteract against this blocking situation will be to consider that:

- Smart grids are not going to be the ultimate evolution of electricity networks. On the contrary, smart grids are a path, rather than a fixed objective.
- No top-down or bottom-up vision has to be used, if it is not done in an iterative way. The solution should come from a balance between realistic and challenging solutions for telecommunications networks.
- Requirements must be realistic and have to lead to achievable expectations in the evolution toward smart grid. On the contrary, requirements just based on simple, inexpensive, and state-of-the-art solutions will just produce a smart grid solution that will be born crippled to support a smart grid vision for years and decades to come.
- A diversity of networks and systems will be needed to solve smart grids today and tomorrow.
- No single solution exists for smart grid ICTs. Focusing on telecommunications networks, and as shown in the other chapters, the difference of requirements needed in different parts of the network, the ubiquity of the solutions to deploy, the limitations in investments and expenditures, and the location of the premises to cover are so different that the use of public and private, wired and wireless, networks will be a must.

Considering the previous aspects, smart grid global solutions composed of application and telecommunications networks must be the result of a unique architecture and system approach. In the case of telecommunications networks, although different technologies and networks are going to be used, there must be a common vision on the way they are going to interconnect and act (for example, unique vision on Ethernet interfaces for the provision of services, common core for all services at backbone level; these kind of approaches will render a solution subject to be operated and monitored in a proper way).

Smart grid evolution and deployment should be planned in phases. These phases will be mandated not only by electricity solutions and appliances that will be deployed in electricity premises, but by the evolution of the telecommunications solutions that are going to be progressively available. And this phase definition, which can be explained in terms of telecommunications service characteristics needed for the evolving needs of electricity systems (such as metering—real time for certain premises, non-real time for others; remote control—low or high bandwidth, real or non-real time; automation; etc.), will also lead to the evolution of the telecommunications technologies and networks that will get to know the use cases of their potential users (utilities), and will prepare themselves to provide what is needed.

In order to be conscious of the implications of these changes, Table 11.2 provides a vision of the situation of the electricity system today, and the future perspective [11].

11.2.1 Purpose of the Smart Grid

The question "Why do we need communications in the electricity grid?" can be answered considering the aspects that will be smart in the grid.

The summarized and pragmatic objectives of smart grids have to do with decreasing consumption with customer-oriented programs (demand side management will act as an indirect source of generation), and improving the efficiency of network operation and maintenance (both to reduce costs and to increase the availability parameters in the grid) through access to intelligent devices inserted in the grid to monitor and act in real time with automated programs.

The objects of smart grids will be the intelligent devices, and the challenge from the ICT perspective is to make them real (electronics), to allow them to be accessed from remote locations (communications), and to make them interact and exchange data cooperating in automatic decision and action taking (applications).

Although this chapter focuses on telecommunications, it is difficult for the noninitiated to separate that specific part of communications from the global system that provides the interface to the services. Thus, it is necessary

Table 11.2 Characteristics of the Network Today and Tomorrow

Today	Tomorrow
Electromechanical	Digital
None or unidirectional communications	Bidirectional communications
Built for centralized generation, with large generation plants	Accommodating distributed generation
Radial topology	Network topology
Few sensors	Monitors and sensors everywhere
None or little monitoring	Self-monitoring
Manual restoration	Semiautomated restoration and, eventually, self-healing
Prone to failures and blackouts	Adaptive protection and islanding
Equipment manual checking	Equipment remote monitoring
Central control	Distributed control
Emergency decisions by group of experts and phone	Decision support systems, predictive reliability
Limited control over power flows	Control systems everywhere
Limited price information	Full price information
Few customer choices	Many customer choices

to mention explicitly that communications will be supporting high-level services for advanced control systems and applications such as:

■ Distribution automation systems, to monitor and perform remote control operations at substations from distant control centers.
■ Energy management system, to make use of data from sensors, meters, etc., to depict the entire system for analysis, control, and planning
■ Intelligent network agents, as the intermediate local point, gathering data and making decisions about local switching and control functions, and communicating results to control centers
■ Demand side management systems, to reduce consumption demand in response to incidents or requests from a utility, whenever it is necessary.
■ Asset management system, to provide a repository for the inventory of facilities and devices

- Geographic information systems (GIS) to map geographic information to electric power, thus providing the correlation of data and its physical position for improved network design, planning, operation, and maintenance.
- Grid modeling, simulation, and design systems to optimize decision taking processes.

11.3 The Different Segments of the Smart Grid

An electricity system is a simple concept. An electricity system is an aggregation of three subsystems, with entity by themselves: generation plants (primary source of the energy) are the core of the generation system; transmission lines (transport of the energy) and substations (to adapt electricity voltage levels for performance purposes—reduction of transportation losses) constitute the transmission system; and the same elements, but with lower-voltage levels, and with a higher number of elements distributed locally, are called the distribution system. To finish with this classification, the endpoint of the distribution system, the points of supply where electricity is delivered, will have a special mention due to reasons explained in other chapters.

In any of the segments, the concept of substation is persistent. A substation is the part of any electricity transmission or distribution network where voltage is transformed from high to medium, medium to low (or vice versa) using transformers. This is a very basic definition of a premise where many other functions may be located (switching transmission or distribution circuits into and out of the grid system, measuring electric power qualities, providing protection from lightning and other electrical surges, etc.).

From a functional perspective, substations can be classified as:

- Generation. Their mission is to transfer electricity from power plants into the network via suitable transformers.
- Transport. These substations work as interconnection nodes of a variable number of transport lines.
- Distribution. Their main role is to reduce the voltage from transmission level to those required for local distribution.

11.3.1 Generation and Transmission System

Electricity is injected in the system at generation assets. This energy is transported to consumption centers with electricity conductors linking substations among them (Figure 11.1).

A primary substation is usually a transport or distribution substation consisting of a set of switching, controlling, and voltage stepping-down or stepping-up equipment, which is prepared to direct electricity, converting

Figure 11.1   Primary substation.

very high voltage (132–400 kV; ranges provided are not exhaustive) or high voltage (45–66 kV) into medium voltage (11–30 kV), or vice versa.

Transformation may take place in several stages and/or at several substations, starting at a power plant substation where the voltage is high for transmission purposes and then progressively reduced to the voltage required for residential, commercial, and industrial loads.

The high voltage (HV) network is normally designed with a ring structure for protection purposes.

11.3.2 Distribution System

A distribution system is probably one of the electricity grid segments that are going to be greatly affected and transformed by the smart grid concepts [12].

A primary substation can feed several secondary substations over medium voltage. Secondary substations can be considered part of the medium-voltage (MV) network. A main component of secondary substrates are MV/LV transformers that transform electricity down to low voltage (LV, <1 kV) customers are usually reached in LV.

The main parts of a secondary substation are:

■ MV lines. These are electricity lines that depart from a primary substation and deliver electricity to secondary substations.
■ Switchgear or MV panels. These are elements with the purpose of isolating and interconnecting transformers and MV lines.

Figure 11.2 Urban underground secondary substation access.

■ Transformers. These are the devices that transform the voltage of a MV distribution line to lower voltages (in the range of 120–480 V). A secondary substation may house one or more transformers.
■ LV busbar. This is an assembly of panels that distribute the electricity from the secondary winding of the transformer to different LV line. Each line usually has fuses or other elements for protection purposes.
■ LV feeders. LV lines support electricity delivered to customers.

Secondary substation characteristics are relevant for their inclusion in the smart grid. They are either indoor or outdoor, depending on where the transformer is placed (Figure 11.2). A distinction can be made depending on the underground or overhead disposition of cables, both MV and LV (as a general rule, overhead in rural (Figure 11.3) and semiurban areas, and underground in urban areas). Different combinations of MV cables, indoor and outdoor placement, and LV cables can be found, depending on the utility and the region.

Equally important is the nature of the switchgear in the MV and LV panels, for the injection of PLC signals where necessary.

The distribution system is composed of both medium-voltage and low-voltage parts of the grid. In both cases, and for the purpose of understanding the possibilities of telecommunications technologies (basically MV PLC) in this segment, it is important to know about the topology of the elements [13].

■ Medium voltage. There are three MV topologies for electricity delivery:

  – Radial topology. Radial topology connects the high voltage (HV) to MV substations with the MV to LV substations (i.e., secondary substation; also referred to as transformer substations) using radial

Figure 11.3   Rural secondary substation.

lines. These lines (also called feeders) can be used for one or more
secondary substrates. Radial systems are the way to maintain centralized control of all the secondary substations. The generalization
of this topology takes the form of a tree-shaped configuration. Radial topologies are in general the easiest and cheapest topology
to develop, operate, and maintain.

— Ring topology. Ring topology appears to overcome the inherent
lack of protection of radial topology capability. A fault in one
element of the MV line may interrupt electricity service in the rest
of the connected substations, leading to the electricity outage in
the area. Thus, a ring topology is an improved radial topology
providing open-link capabilities to other MV lines to allow for
the creation of redundancies. The grid is always operated radially
from a HV to MV substation, but in the event of a faulty section
of the feeder, other elements switch to reconfigure the grid such
that most faulted lines are quickly restored.

— Networked topology. Networked topology means that substations
are connected through multiple MV feeders, in such a way that
the reconfiguration capabilities to overcome faults are multiple. In
the event of failure, multiple solutions can be found to reroute
electricity supply.

■ Low voltage. LV networks have relatively more complex and mixed
topologies than MV networks. There are multiple causes for this:

Table 11.3   Typical Data for Electricity Grid in Europe

	High-Density Residential Area	Low-Density Residential Area
Type of secondary substation (SS)	Underground or on ground level inside a building	House type or over a pole
Transformers per SS	2	1
Average number of customers per SS	250–320	100 (10–200)
Low-voltage (LV) feeders per transformer	6–8	6–8
Average length of the LV lines (m)	150	300 (100–800)
Type of LV line	Underground	Overhead
Customers per metering room	10–25	1–4

extension and characteristics of service area, the "loads" (points of supply), country- and utility-specific criteria and operating rules, etc. A transformer in a secondary substation typically feeds several LV lines, with one or more MV to LV transformers. Topology is often radial with a tendency to create tree-shaped lines; however, there are also networked grids, and ring configurations found in LV networks. LV lines are generally shorter than MV lines, their characteristics change depending on the service area, as reflected in Table 11.3 [14].

11.3.3 Points of Supply

Points of supply are, strictly speaking, part of the distribution system. The points of supply are the ends of the electricity system, nearer to customer consumption. For this reason, they create a very distributed structure, present in the homes of the consumers, in a number similar to the number of households and industries. If we geographically locate all the points of supply, the picture we get is representative of populated areas.

The points of supply are a very relevant part of the system that has often been neglected. From an electricity system point of view, and in a not very competitive market, subscribers were paid attention just to be charged. For this purpose, points of supply are equipped with meters that accumulate consumption, and once in a while, this consumption is annotated for billing purposes.

It has not traditionally been a pressure for utilities to make a big issue from metering business. The perspective was simple: some people are sent to the buildings to read meters, trying not to spend much money on this effort, because if the reading cannot be taken today, it will be done some other day. In the end, all the consumption will be billed, because the meter is consistently and reliably accumulating consumed energy.

But metering business is no longer so straightforward. Liberalization and competitiveness are making this situation change. Even in some countries, the metering business has been taken apart from the distribution business itself, considering it one element behind the liberalization effort of the market, even apart from the energy commercialization business.

But even if regulation is not making the metering business change, there are other forces that favor changes in this sector, among them, fraud detection, remote connect/disconnect, technical losses, effective control, and new services such as demand side management. All these elements require telecommunication capability in the meters associated to the points of supply.

A meter is an electricity measuring instrument that can be found at the points of supply, and that totalizes the electric energy consumed by users over a given period. Meters have a different ownership strategy in the different electricity systems in the world. In some cases, they are the property of the customers; in others, of the distribution company; and still others, there are third companies in charge of the metering business itself. The common factor in all of the situations is that the meter is at the border of the distribution grid to measure energy consumption.

Meters have traditionally been electromechanic devices. Since the advent of the smart grid initiatives there is a trend toward the massive substitution of these devices by electronic meters, [15,16] in some cases with improved functionalities, but all of them with the objective of being remotely accessed. Telecommunications are needed for this purpose.

Meters are associated to the customers, and their placement greatly depends on the nature of the customer premises. Many meters have a distributed location, so that only one meter can be found at a certain location; other meters are placed at shared rooms in the basement of buildings, all together, facilitating the process of meter reading, either manually or automatically, with automatic meter reading (AMR) systems.

11.4 Communications in the Different Segments of the Smart Grid

Focusing on the communications component of ICTs, these smart devices in smart grids require the ability to communicate. The specific communication media over which they share information is not relevant, as long as it fulfills certain conditions related to throughput, delay, security, privacy, etc.

Figure 11.4 Carrier's telecommunications network supported on primary substation assets.

There are many aspects that have to be considered when using telecommunications networks for smart grids (Figure 11.4). Following are some of the important factors to be remembered:

- Design of an integrated and structured telecommunications network for the different services needed.
- Future-proof nature of the decisions taken over the investments in telecommunications networks. Technologies in telecommunications often have a limited life span that is out of the decision capabilities of utilities. Thus, having to replace a technology by another one in a short period of time will affect business plans, because installation costs are many times higher than investments in equipment and terminals.
- Availability and viability of telecommunications networks and services in the scenarios handled by utilities. Environmental conditions, pole mounting, and underground placing are not usual service locations for telecommunications services.
- Balance between public and private assets and networks. In the development of a network, capital expenditures and operational costs must be balanced to achieve a cost-effective solution for the network. Incurring huge investments is probably as bad as allowing public networks

to be the unique source of solutions for a network. Additionally, as has been the case in the past with many carriers' initiatives, the possibility of joint efforts between carriers and utilities should be considered. A regular telecommunications carrier may take advantage of the rights-of-way of utilities, and these utilities may advantageously use the experience of carriers in the management of complex telecommunications networks.

11.4.1 General Considerations

Electricity networks impose severe constraints [17] on the telecommunications services that are needed for their efficient operation and maintenance.

Electricity premises are not regular telecommunications carrier sites, and although they share certain similarities with industrial sites and, in some cases, with residential sites, they are neither one nor the other. This basic fact is going to affect the conditions of the devices that can be installed in them, either when the telecommunications devices are placed there in order to create private networks, or as terminals of public networks.

Telecommunications devices have to comply with strict standards imposed by the characteristics of the electricity devices placed in this environment [18–21]. Table 11.4 summarizes the different aspects to take into account when the decision is made on the acquisition or design of a new telecommunications device for an electricity premise.

Apart from the device specific considerations of the table, some more aspects may be detailed, regarding different aspects of the premises to cover:

■ Location. Utility premises pose a challenge when we refer to distribution assets. If we refer to secondary substations, we can find a classification of pole-mounted transformers, underground enclosures, surface shelters, basement prepared spaces, etc. All of these premises have different difficulties as part of a telecommunications network. Pole mounting has difficulties typical of outdoor and tower mounting configurations; underground premises are not easily reached with radio communications, and may experience floods in certain periods of the year; surface shelters often have space constraints, and it is difficult to find room for telecommunications devices; etc. In all cases, physical access to such premises is also a challenge, because many of them were designed not to be visited often.
■ Reliability of power supply. The deployment of DC batteries along with the corresponding rectifiers (which, by the way, occupy additional space and are subject to temperature conditions described previously) is needed if the service is to be maintained when electricity outages occur. This is also a condition for any public technology

giving support to smart grid deployments, because it often happens that the telecommunications infrastructure of the carrier relies on the electricity supply of a secondary substation, which at the same time is using the telecommunications services provided by this infrastructure. Whenever there is a failure in the substation, the telecommunications infrastructure is going to be affected, and the telecommunications service will eventually become nonoperative.

11.4.1.1 Quantitative Measure of the Services Needed

The premises of the electricity grid where telecommunications services for smart grid are required are:

■ Substations, generation and transport. These infrastructures are usually core facilities located far from city centers. In these places, there is no space constraint, and the investments involved are big enough for telecommunications not to be a problem. However, these premises are often not easily accessible, and telecommunications are not always provided by public carriers. Telecommunication requirements must be considered since the beginning, in order to have the investments well identified in the overall budgets.

■ Substations, distribution.
  − Primary level. These substations are close inside or to cities (there might be several in a single city, initially placed in the outskirts, but eventually absorbed in the expanding center), and although they are smaller than generation and transport substations, telecommunications are not a major concern at the time they are built.
  − Secondary level. These substations are very diverse in nature. At the time most of them were built, telecommunications were not needed there. Pole, surface-mounted, or underground transformers are found in these premises, in urban, suburban, and rural areas. The size of these premises tends to be smaller with each new generation design. Much of the challenge of telecommunications for smart grids is placed here.

■ Fuse boxes, meter cabinets, and meter rooms (Figures 11.5 & 11.6). Fuses to limit responsibility or for maintenance purposes are found near individual meters or concentrations of these. The enclosures used to place both elements are often tightly fitted to the size of the element itself, and the possibilities for telecommunications are restricted to small devices or integrated solutions inside the elements themselves. An example of integrated telecommunications is the integration of communications modems with electricity meters.

Figure 11.5   Fuse boxes (external appearance).

The challenge of smart grids is primarily placed in secondary substation and meter needed services. The rest of the areas are usually well covered from the telecommunications perspective, and if they are not, the importance of their assets and the relevance of the investments there, justify the deployment of a telecommunications solution based on fixed-network telecommunications assets (fiber optics, radiolinks, etc.).

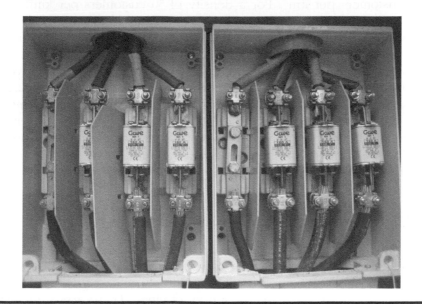

Figure 11.6   Fuse boxes (internal appearance).

Figure 11.7 Concentration of smart meters.

To have a rough estimate of the quantity of electricity premises, it is important first to focus on a geographical area. Even with this focus, we will not be able to be very specific, because figures vary greatly depending on the country, and inside the country, and even in the utility in that country if there is more than one.

As an example [13], in Europe population density varies from 1.4 to 1,400 customers per km². For a density of 50 customers per km², there could be a number of 7,200 customers per primary substation. At the secondary substation level, we could find around 60 secondary substations for each transport or primary substation, and 1 secondary substation per 120 customers. Regarding concentration of meters (Figure 11.7) in meter rooms (and this is important in terms of telecommunications technologies [22,23]), first we have to know if the utility has a trend to concentrate such elements; if not, individual meters are going to be found in each of the households individually. If the utility concentrates meters, the average number of households per building should be considered. A typical urban figure could be 30 meters per building, and 15 for suburban areas.

11.4.2 Generation and Transmission System

Electricity networks today carry energy from remotely located production facilities to consumption centers, via transport and distribution networks. Transport facilities are rarely located inside cities, and primary distribution assets are usually in the periphery of cities. These assets are high-importance assets, because they concentrate much of the responsibility of

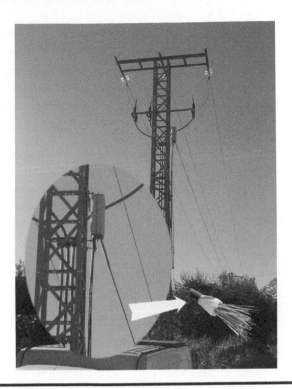

Figure 11.8   Fiber optic cables over electricity poles.

the service for big groups of the population. Utilities have traditionally used their own private telecommunications networks to provide telecommunications service for this environment. The reason for this has to be found in the history of telecommunications, and in the difficulties to combine interests of different companies. Telecommunications companies create telecommunications networks where they can obtain a profit from them (Figure 11.8). Thus, rural areas and low-population villages have not traditionally been the target of their networks. However, electricity generation sites tend to be in such places. Utilities were forced to provide the telecommunications they needed for these assets at their own risk and investment, and began a trend that reaches our days. The service and quality level provided by these privately owned networks in many cases exceeds that of public networks, and in any case it is controlled by the utility itself.

Private telecommunications networks for utilities are a coordinated group of assets and technologies that provide a controlled and reliable service over the premises they serve. Specifically, the network is made of a physically supporting network (fiber optics, radio, PLC), with transmission technologies placed over those (TDM, SDH/SONET, WDM). Working over these transport layers, voice and data switching (frame relay, MPLS, Ethernet) solutions provide the final service for applications:

- Physically supporting network. Utilities often place fiber optic cables over their electricity poles. Optical ground wire (OPGW) and all dielectric self-supporting (ADSS) cables are industry examples of such cables, and the kilometers of fiber optic cables in utilities are supported on a high percentage of their electricity lines. As a complement to those, utilities tend to create microwave radiolinks to provide backbones and radio access where they cannot place fiber, or where they need a backup for those. Long distance power line (km) is an old telecommunications asset that utilities tended to use in transmission systems mostly in the past, and that is still alive in other segments of the network. However, PLC technologies are taking an active role in the configuration of the smart grid.
- Transmission technologies. Time division multiplexing (TDM) has evolved toward synchronous digital hierarchy (SDH)/SONET (synchronus optical network) technologies, and these last ones are being grouped in wave length division (WDM) links. As in any other business, the trend is supported in the need to cope with bandwidth-hungry applications, whose data transport demand is increasingly bigger.
- Switching technologies. "Old" TDM channels (64 kbps, 2 Mbps) are being replaced by Ethernet interfaces (10/100 Mbps; 1 Gbps), switched over multiple label switching (MPLS) cores, with either layer 2 or layer 3 versions of this technology.

As a conclusion, we can say that there will not be an application in the near future that poses a challenge for the telecommunications assets existing in these generation and transmission environments, where the usage of telecommunications infrastructure and means is well known and controlled by utilities.

11.4.3 Distribution System and Points of Supply

As it has been described in general terms, the assets involved in this part of the network are distributed in the service territory of a utility, and both their distributed nature and their number (exceeding traditional generation and transmission system figures) make the telecommunications service in this part of the grid the real challenge of the smart grid environment.

11.4.3.1 Medium-Voltage (MV) Grid

If we take the previous chapter as a reference, we will try to identify the assets we could take advantage of, if we were to use the same private network model:

- Power line poles can support telecommunication cables (fiber optic), and they also exist in this segment of the network. ADSS will probably be the most prevalent model.
- Power line poles can also support antennas for radio/wireless telecommunication.
- Power line cables can be used as the conducting media for telecommunication signals with a reuse of the infrastructure that is similar to the one provided by xDigital Subscriber Line (xDSL) technologies over traditional telephony copper pairs.
- Ducts, in which underground power lines are installed, are also valuable infrastructure electricity companies may use to install fiber optic underground cables.
- Secondary substations, which are located inclose proximity to end customers, often very house hybrid fiber—coaxial and fiber—carriers [24,25].

Although this mentioned infrastructure often helps to build a private network for the utility, smart grids are going to require a change in the traditional services provided, toward pervasive reach and broadband networks, with severe latency, security, and reliability constraints, even if these last performance parameters are difficult to achieve (especially with public-oriented telecommunications networks).

This paradigm change makes us realize that a blended approach of utility and public telecommunications carrier services might be needed, at least in the short term, where the ability and investment capacity of the utility will be handicapped by the need of quick and "inexpensive" roll-outs. In this realistic approach of combining both public and private networks, PLC will play a central role.

11.4.3.1.1 PLC over Medium-Voltage (MV) Grid

According to the data provided, it is common to find one secondary substation for every few 10 or 100 customers. Thus, the number of this type of substations is important, and the telecommunications technology to be used should be selected with care. Considering the nature of secondary substations, it can be easily understood that they tend not to be easily accessible, from both the physical access and telecommunication points of view. We can find secondary substations on top of poles in rural areas, and underground in densely populated areas. The situation is that what is suitable for a certain kind of secondary substation (radio, for example) might not be suitable for the other.

The specific telecommunications asset that is always available at a secondary substation is power line. Thus, PLC technology can be applied to secondary substations, for both overhead and underground placements.

The first question to decide is to use either broadband PLC or narrowband PLC. If we take into consideration future and evolving needs of smart grids, the trend would be to use broadband PLC (also known as BPL [26,27]); although if we place our attention on what is fundamental as a telecommunications service, we would conclude that narrowband PLC can be enough. Possibly the best strategy would be to use BPL in MV, to making the best use of the conceptual split of the LV grid (where narrowband PLC seems unavoidable for simplicity reasons), and leave BPL for the MV segment. It has to be added that BPL will be used to configure an interconnected network among secondary substations, and that this secondary substation local network will be connected to external telecommunications networks by means of backbone connections of different technologies.

BPL has a long recent history of usage in the electricity grid. Recently, BPL has been extensively used in home environments, to overcome radio technology limitations for in-home services. And here, there is another reason to make use of BPL in MV: we are far from in-home environments, and no interference will be caused by parallel deployments.

The deployment of BPL technology, although very useful for utilities due to its inherent relationship with power lines, requires procedures for device installation, service provisioning, and network operation and maintenance, closely related with electricity grid operation. A utility must review all the existing procedures and activities directly involved with the installation of devices in electrical locations to ensure that BPL deployment is not going to be affected by the electricity business itself, and the other way around.

Telecommunications signals must be coupled to the MV lines. This signal injection is achieved with coupling installation, which is probably the most critical task from an electricity distribution point of view, since couplers are intrusive elements in the grid that must be installed under very restrictive operational procedures that involve periods of outage in the grid. Even more, coupling device manufacturers have to test and approve their products with regard to electrical safety standards. Depending on the type of power line MV cable and MV cabinet present at the secondary substation, it is necessary to select the type of coupler needed among a set of capacitive and inductive varieties, which must be tolerant to different electricity grid maneuvers (Figure 11.9). The installation and handling of couplers must be done directly by utility authorized staff (even access to the premises is only allowed for specifically qualified people). Special attention must be paid to the different positions that the switches of the different substations might have; in all of these positions, couplers must continue working in the PLC range.

The most important aspect of MV PLC deployment is the planning phase. There are many PLC technologies available for MV, and a few of them are among those selected and developed by IEEE and ITU. The deployment of a PLC network in MV is not a straightforward activity because, apart from

Figure 11.9 Medium-voltage capacitive coupler.

the connectivity-specific aspects (see couplers), PLC is a self-interfering technology. This circumstance means that one PLC link may interfere with another if no provisions have been taken to avoid this circumstance, and in any case, these provisions are surely going to reduce global performance of the PLC network.

If PLC technology offers a possibility to deploy a network, disregarding possible interference among the different links, the issue to consider is capacity. As the technology in its maturity requires less attention to the PLC link constitution, efforts will be placed in ensuring a certain performance once the whole network is deployed. This effect will be achieved by placing backbone connections in certain secondary substations, with the intention of limiting the layer 2 network extension, to control performance in each layer 2 domain.

If PLC technology offers the possibility of planning networks to avoid interferences, attention should be paid to avoid interferences among links sharing the same resources, and provisions should be taken to guarantee interference-free domains, based in the knowledge of the electricity networks, and as many conservative rules as needed to count for future network expansion.

In both cases, several statements can be compiled as a summary:

■ For MV PLC network deployments, a good knowledge of electricity network distribution and composition is needed.
■ MV PLC solutions will be extended over many secondary substations, but it is highly likely that not all secondary substations can be part of the PLC solution.
■ Performance (simultaneous bits per second in the network) is the factor to consider when planning the extension of the MV PLC network, and the number of external backbone connections to deploy.

■ MV PLC uses power line infrastructure to communicate. PLC connections will have to be guaranteed even if switches in the lines are opened, and even if a certain power line is not working. Alternative communication paths are needed in the network.

11.4.3.2 Low-Voltage (LV) Grid

A low-voltage network is important in smart grids due to the access to the final consumption points.

By means of this access, and referring to metering, utilities can understand the amount of energy extracted from the network, compare it with the amount flowing out of secondary substations, and understand what and where their technical losses are (energy lost in LV distribution grid). Apart from technical losses, and thanks to the ability of electronic meters to store consumption patterns, utilities will be able to understand the different customer types they have, and how their behavior could improve global energy delivery; utilities, or the traders and retailers of energy, will be able to adjust their energy offers in real time; customers will be able to make decisions on their energy consumption, take advantage of those offers, and the system as a whole will be optimized.

Equally important to the traditional metering is the ability to communicate in real time with these borders of the network. Modern meters include the ability to connect/disconnect. Thus, in events such as fraud detection, demand side management programs, or new service activation, the utility will be able to remotely operate the meter.

But placing communications in meters is not straightforward. Several aspects must be considered to understand this statement:

■ The quantity of meters is huge. It can be estimated that the number of existing meters is approximately equal to the number of households. These figures are a challenge to any existing telecommunications network, due to the fact that it probably implies making its customer capacity double.

■ Real-time capabilities and low-capacity communication capabilities are requirements of telecommunications networks for these services. The first requirement is demanding for massive access telecommunications technologies; the second is not very attractive for a carrier taking the risk of investing to cope with these new entrants in the network.

■ Meters are placed in locations where access to telecommunications is not immediate. Basements of buildings in urban areas and isolated locations in rural areas are representative examples.

As it has been commented in other sections of this chapter, telecommunications technologies for smart grid are not going to be unique. On the

Figure 11.10 **Urban secondary substation. Private radio telecommunication access.**

contrary, utilities have assumed the need to combine several, in both the different segments and each particular segment. The following are options we can find today:

■ Radio technologies. Radio alternatives are attractive, due to their theoretical wide penetration (Figures 11.10 and 11.11). There are two types of these technologies:

Figure 11.11 **Urban secondary substation. Public radio telecommunication access.**

- Public networks. Existing public networks provide data communication capabilities, 2G technologies such as GPRS and 3G technologies such as UMTS.[28,29]
- Private networks. There are license- and non-license-oriented radio frequency bands, over which standard or proprietary systems can be deployed. Examples of these are WiMAX type
  technologies.[30–32]
- Wired technologies. Cables are a telecommunications assets prepared
  for the future in the sense that, as a basic infrastructure, the cables can
  be reused whenever there is a new technology capable of extracting
  more capacity from it.
  - Public networks. xDSL [33] technologies and cable modem tech-
    nologies are capable of providing the type of service needed.
  - Private networks. Although many utilities own copper pairs and
    fiber optic cables, these elements are not usually present where
    meters are located. However, there is an asset always present
    where meters are located, and that utilities have traditionally been
    using for their own private network: power line.

Many utilities are following a strategy that tends to minimize the usage
of public networks, as long as it can be avoided when the balance of
investments and operational expenditures is adequate. Thus, they tend to
think of private networks for their massive roll-outs of communications in
LV, to complete areas not covered by public networks.

When private networks are involved, the choice between radio and
wired technologies (mainly PLC) is basically made depending on the place
where meters are located and the nature of their concentration. If meters
are placed in open air, with low densities, there is a trend toward selecting
radio (Figure 11.12), and if we have meters inside buildings in difficult radio
access zones, the choice is PLC.

Both radio and PLC technologies used for this purpose are "last mile"
technologies, and both need to be supported and complemented with other
technologies capable of concentrating their accesses, and working in the
long-distance transmission. The premises where utilities concentrate these
last-mile technologies are generally substations.

11.4.3.2.1 PLC over Low-Voltage (LV) Grid

The most common trend when using LV PLC technologies to access meters
is to use narrowband PLC, [34–42] especially in Europe. The reason behind
this is that the biggest utilities pursuing smart grid and/or smart metering
deployments are focused on their own operational needs, and they do not
envision any other usage for these customer accesses than plain operational
purposes. Even more, the cost of broadband PLC technologies [43] and

Figure 11.12 Rural secondary substation. Public radio telecommunication access.

greater difficulties in their deployment usually prevent utilities from using broadband for massive remote meter access.

Another important aspect of narrowband PLC technologies is that many of them were developed decades ago, [44,45] and their characteristics are no longer state of the art—simple FSK modulations, low capacities, big latencies, no repetition capabilities, no plug-and-play deployment style, etc.

The adaptation of broadband PLC concepts to narrowband PLC technologies has made some new PLC technologies appear in the market. Standardization bodies (IEEE, ITU [46,47] have also placed some attention on these efforts and are trying to help in the process of making standards a reality. However, utilities have the urgent need to deploy AMR systems as the first step toward their smart grid, and what is really a value for utilities is to have a sound technology, working smoothly in the field, and with such openness that multiple supply sources exist to provide meters and substation concentrators for them to deploy. One such example is PRIME technology, which is attracting attention all through the world, due to its openness (no patents or royalties applicable), market support (PRIME Alliance gathers tens of stakeholders, creating the proper ecosystem), state-of-the-art techniques (OFDM multiplexing, different adaptive modulation schemes and error correction mechanisms, sound layer 2 MAC, etc.), and above all, the real plug and play behavior in the field, which allows the deployment of meters to be a completely electric business.

The Powerline Intelligent Metering Evolution (PRIME) [48–53] project was launched in 2006 to provide open, royalty-and patent-free, physical and media access control layers, along with the definition of certain convergence layers (IP, DLMS/COSEM, etc.), for narrowband PLC solutions in the CENELEC-A band [53] (3 to 95 kHz) in the low-voltage distribution network. Maximum throughput is 120 kbps.

PRIME technology is designed to provide a telecommunications network over PLC to connect meters to their secondary substation. At secondary substations other communication technologies are used provide links to the central systems.

One of the most outstanding characteristics of PRIME is its MAC layer. The MAC layer takes some ideas from meshed systems, but considering that the available bandwidth will be limited it provide singular characteristis. With CSMA/CA as the access method and the concept of "switch" as (a meter helping the base node) i.e. the concentrator to create the network at the secondary substation) to reach every part of the network, it is capable of creating a structure without any human intervention in any of the network devices.

Plug-and-play mechanisms must not hide planning aspects of any technology deployment. Plug and play means that no or little configuration efforts are needed in the system. However, when a PLC technology is going to be deployed in a utility, the telecommunications technology is so tied to the asset that the deployment strategy must count on the details of the asset, to be able to control the results obtained.

PLC technologies have several limitations inherent to their nature:

- Distance. As in any telecommunications technology, distance affects the propagation of PLC signals. Because PLC is a last-mile technology, we can expect the distance range to vary between a few hundred meters to a few kilometers. This distance will be different, depending on the nature of the electricity cable and its deployment style, (in general, overhead lines will propagate signals better than underground cables).
- Number of system elements. PLC communications channels is shared. All elements inside this PLC medium share the same communication channel, and performance will decrease in accordance with the number of elements working at the same time and their concentration in a location. Luckily, AMR systems are today master-slave oriented, and what happens is that the master (concentrator at substations) polls slave (meters), avoiding too many simultaneous transmissions having said this, modern PLC technologies, especially if they exhibit plug and play characteristics, offer a MAC that has a group of control messages that work in the background consuming the resources needed for the maintenance of the network, and whose simultaneous operation must be guaranteed with the proper performance.
- Concentration of elements in the same location. The concentration of PLC elements in a location leads to situations in which many devices share a collision domain. Even if the system is master-slave, not all signaling (control packets) can be controlled, and the performance will also be different this element is taken as a parameter.

■ Noise. This factor is implicit in any communication system, and especially for PLC is a very restrictive factor. Noise is the enemy against which the entire PLC system has been designed (especially at the physical layer level) to offer the best of its possibilities.

Of all the elements mentioned above, such as distance and the number of elements can be characterized in the network of a utility. The rest are just constraints that we will find decreasing and conditioning the performance of the network; their occurrence will also be statistical, in both the location and time aspects (there are studies on the noises, and some general aspects can be clearly fixed: general level lower at night than during the day, more likely in dense areas that in disperse areas, etc.).

Thus, independently of the specific PLC technology under use, LV grids can be classified around the parameters of distance (of the low-voltage lines), number of meters in the lines fed by the substation, and concentration of meters in the same meter room. It should be expected that the grouping of substations around those factors with similar value ranges should statistically provide similar performance results.

11.4.4 Distributed Generation

Distributed generation is evolving and making the whole electricity generation paradigm change. Distributed generation solutions are making the center of energy production come closer to end customers, and in this process, it will often be the case that we will find these energy sources mixed with traditional consumption points.

From a telecommunications perspective, the nature of the solutions needed for this generation will share many aspects with the other transport and distribution assets present in the same places. Thus, depending on the state of the art and the existence of public networks or utility private networks, different options will be present. The important aspect to consider is that from a telecommunications perspective, these options will not greatly differ from the ones presented in previous chapters.

11.5 Conclusion

Smart grid implementations are paving the way for smart grid visions to happen. Real infrastructure conditions the way in which smart grids are taking shape, but more important than the starting point of the implementations is the strategy we use to achieve the goals these visions imply.

This chapter focused on providing insights over the existing possibilities, in terms of technologies, networks, and systems in the telecommunication component applicable to smart grids. The chapter outlines the strategy to be used when designing a telecommunications network for smart grids, taking into consideration the constraints existing in the networks nowadays, and the fact that smart grids are a design philosophy rather than a specific existing solution to implement.

References

1. UTC Research Report, *Smart Grid Economics. Making the Business Case for Smart Network Technology* (2009).
2. Office of Electricity Delivery and Energy Reliability, *The Smart Grid: An introduction*, prepared for the U.S. Department of Energy by Litos Strategic Communication, December 2011: http://energy.gov/oe/downloads/smart-grid-introduction-0 (2008).
3. Global Environment Fund/Center for Smart Energy, *The Emerging Smart Grid: Investment and Entrepreneurial Potential in the Electric Power Grid of the Future Center for Smart Energy* (2005).
4. Global Environment Fund/Global Smart Energy, *The Electricity Economy. New Opportunities from the Transformation of the Electric Power Sector* (2008).
5. H. Yan, N. Jenkins, W. Jianzhong, and M. Eltayeb, *ICT Infrastructure for Smart Distribution Networks*, in *ISPLC10*, Rio de Janeiro, Brasil, 2010, pp. 319–324.
6. R. Mora, A. Lopez, D. Roman, A. Sendin, and I. Berganza, *Communications Architecture of Smart Grids to Manage the Electrical Demand*, in *WSPLC09*, Udine, Italy, 2009.
7. V.K. Sood, D. Fischer, J.M. Eklund, and T. Brown, *Developing a Communication Infrastructure for the Smart Grid*, in *IEEE Electrical Power & Energy Conference (EPEC) 2009*, Montreal, 2009, pp. 1–7.
8. U.S. National Broadband Plan, *Connecting America, the National Broadband Plan*, http://www.broadband.gov/. Last accessed December 2011.
9. ITU, *Focus Group on Smart Grids*, http://www.itu.int/en/ITU-T/focusgroups/smart/Pages/Default.aspx. Last accessed December 2011.
10. 101-IEEE, *Smart Grid*, http://smartgrid.ieee.org/. Last accessed December 2011.
11. European Commission, *European SmartGrids Technology Platform: Vision and Strategy for Europe's Electricity Networks of the Future*, http://ec.europa.eu/research/energy/pdf/smartgrids_en.pdf. Last accessed December 2011.
12. Distribution Study Committee, *Distribution Network Configuration and Design: Network Design—Applied Practices in European Countries* Union of the Electricity Industry—EURELECTRIC, 50.04.DISNET, Ref. 05004, Ren 9539, 1995.
13. H.C. Ferreira, L. Lampe, J. Newbury, and T.G. Swart (editors), *Power Line Communications: Theory and Applications for Narrowband and Broadband Communications over Power Lines*, Wiley, New York, 2010.
14. Open PLC European Research Alliance (IST Integrated Project 507667), *Report Presenting the Architecture of PLC System, the Electricity Network Topologies,*

the Operating Modes and the Equipment over Which PLC Access System Will Be Installed, http://www.ist-opera.org.

15. I. Berganza, *Iberdrola Strategy on AMI and Smart Grids, Metering International, Issue* 2, 2008.

16. Energy Retail Association, *Smart Meters*, http://www.energy-retail.org.uk/smartmeters.html. Last accessed December 2011.

17. Iberdrola, *Technical Manuals and Norms*, Iberdrola Distribución, Spain, 2011.

18. A. Sendín, R. Guerrero, P. Angueira, and J. Morgade, *PRIME Technology in Smart Electricity Grid. Adaptation of Telecommunication Systems to Electricity Grid*, in *49th FITCE10*, Santiago de Compostela, Spain, 2010.

19. A. Sendin, A. Llano, A. Arzuaga, and I. Berganza, *Field Techniques to Overcome Aggressive Noise Situations in PLC Networks*, in *IEEE ISPLC11*, Udine, Italy, 2011.

20. A. Sendin, A. Llano, A. Arzuaga, and I. Berganza, *Strategies for PLC Signal Injection in Electricity Distribution Grid Transformers*, in *IEEE ISPLC11*, Udine, Italy, 2011.

21. Schneider Electric, *Electrical Installation Guide*, http://www.schneider-electric.com. Last accessed December 2011.

22. IEEE 802.15.4, *Wireless Personal Area Networks (WPANs)* (2006). http://standards.ieee.org/about/get/802/802.15.html. Last accessed December 2011.

23. ZigBee Alliance *ZigBee Specification*, http://www.zigbee.org/Standards/Overview.aspx Last accessed December 2011.

24. Docsis, *Data over Cable Service Interface Specifications*, http://docsis.org/, http://www.cablelabs.com/specifications/ (2011).

25. FFTH Council, *Fiber-to-the-Home*, http://www.ftthcouncil.org/. Last accessed December 2011.

26. HomePlug Alliance, *HomePlug Specifications*, http://www.homeplug.org. Last accessed December 2011.

27. IEEE 1901, *Broadband over Power Line Networks: Medium Access Control and Physical Layer Specifications*, http://grouper.ieee.org/groups/1901/. Last accessed December 2011.

28. A. Sendin, *Foundations of Mobile Communication Systems. Evolution and Technologies* (McGraw-Hill, New York, 2004).

29. 3GPP, *3rd Generation Partnership* http://www.3gpp.org. Last accessed December 2011.

30. WiMAX Forum, *WiMAX Case Studies* http://www.wimaxforum.org/ (2011).

31. Wi-Fi Alliance, *Wi-Fi for the Smart Grid*, http://www.wi-fi.org/. Last accessed December 2011.

32. IEEE 802.20, *Mobile Broadband Wireless Access (MBWA)*, http://www.ieee802.org/20/. Last accessed December 2011.

33. DSL Forum, *DSL Technology Evolution*, http://www.broadband-forum.org/. Last accessed December 2011.

34. A. Zaballos, A. Vallejo, M. Majoral, and J.M. Selga, *Survey and Performance Comparison of AMR over PLC Standards, IEEE Transactions on Power Delivery*, 24 (2), 604–613, 2009.

35. D. Sabolic, *Influence of the Transmission Medium Quality on the Automatic Meter Reading System Capacity, IEEE Transactions on Power Delivery*, 18(3) 725–728, 2003.

36. K. Honghai and W. Zhengqiu, *Application of AMR Based on Powerline Communication in Outage Management System*, in *SUPERGEN09, International Conference on Sustainable Power Generation and Supply*, 2009, pp. 1–4.

37. B.S. Park, D.H. Hyun, and S.K. Cho, *Implementation of AMR System Using Power Line Communication*, in *IEEE/PES Transmission and Distribution Conference and Exhibition 2002: Asia Pacific*, 2002, Vol. 1, pp. 18–21.

38. C. Nunn, P.M. Moore, and P.N. Williams, *Remote Meter Reading and Control Using High-Performance PLC Communications over the Low Voltage and Medium Voltage Distribution Networks*, in *7th International Conference on Metering Apparatus and Tariffs for Electricity Supply*, 1992, pp. 304–308.

39. P. Oksa, M. Soini, L. Sydanheimo, and M. Kivikoski, *Considerations of Using Power Line Communication in the AMR System*, in *IEEE ISPLC06*, 2006 pp. 208–211.

40. A. Hosemann, *PLC Applications in Low Voltage Distribution Networks*, in *ISPLC97*, Essen, Germany, 1997, pp. 134–139.

41. I. Berganza, A. Sendin, and J. Arriola, *PRIME: Powerline Intelligent Metering Evolution*, in *SmartGrids for Distribution, IET-CIRED*, CIRED Seminar, 2008, pp. 1–3.

42. A. Lotito, R. Fiorelli D. Arrigo, and R. Cappelletti, *A Complete Narrow-Band Power Line Communication Node for AMR*, in *IEEE ISPLC07*, 2007, pp. 161–166.

43. L. Weilin, H. Widmer, and P. Raffin, *Broadband PLC Access Systems and Field Deployment in European Power Line Networks IEEE Communications Magazine*, 2003, 41(5), 114–118.

44. D. Dzung, I. Berganza, and A. Sendin, *Evolution of Powerline Communications for Smart Distribution: From Ripple Control to OFDM*, in *IEEE ISPLC11*, Udine, Italy, 2011.

45. N. Pavlidou, A.J. Han Vinck, J. Yazdani, B. Honary, *Power Line Communications: State of the Art and Future Trends, IEEE Communications Magazine*, April, 2003. 34–40.

46. IEEE Low-Frequency Narrow-Band Power Line Communications, *Working Group*. Last accessed December 2011. http://grouper.ieee.org/groups/1901/2/ (2011).

47. ITU, *G.hnem Home Networking and Energy Management*, http://www.itu.int/ITU-T/studygroups/com15/index.asp. Last accessed December 2011.

48. A. Sendin, *Iberdrola PRIME Project: Above PHY Layer, Metering International*, 4, 2008.

49. A. Sanz, J.I. Garcia-Nicolas, P. Estopian, and S. Miguel, *PRIME from the Definition to a SoC Solution*, in *IEEE ISPLC09*, 2009, pp. 347–352.

50. I.l. Han Kim, B. Varadarajan, and A. Dabak, *Performance Analysis and Enhancements of Narrowband OFDM Powerline Communication Systems*, in *First IEEE International Conference on Smart Grid Communications, SmartGridComm*, 2010, pp. 362–367.

51. PRIME Alliance—Technical Working Group, *PRIME Specification*, version 1.3e, May 2010, http://www.prime-alliance.org (2010).

52. A. Arzuaga, I. Berganza, A. Sendin, M. Sharma, and B. Varadarajan, *PRIME Interoperability Tests and Results from Field*, in *IEEE Smartgrid Conference*, Gaithersburg, MD, (2010).

53. PRIME Alliance, *PRIME Specifications*, http://www.iberdrola.com/suppliers/SmartMetering/. Last accessed December 2011.

54. EN 50065-1, *Signaling on Low-Voltage Electrical Installations in the Frequency Range 3 kHz to 148.5 kHz—Part 1: General Requirements, Frequency Bands and Electromagnetic Disturbances.*

Chapter 12

Study on ICT System Engineering Trends for Regional Energy Marketplaces Supporting Electric Mobility

Christian Müller, Jens Schmutzler,
and Christian Wietfeld

Contents

Integrating spatially decentralized renewable energy resources comes along with new requirements for underlying ICT infrastructures in upcoming smart grids. In order to coordinate and balance energy supplies and demands under these new circumstances, various system level challenges for regional power distribution need to be addressed. Hence, this chapter introduces a reference system architecture for *regional energy marketplaces*. The architecture design is based on the overall idea that using distributed and renewable energy resources requires direct involvement of end consumers in order to allow efficient power distribution and load management. In such a scenario, regional marketplaces become central data hubs and mediate demands and supplies between all marketplace participants and stakeholders. This work discusses current state of the art for connecting end consumers to such platforms. Defining a set of system level interfaces respects potentially upcoming requirements, new evolving services, and—resulting from these services—integration and mediation of future market roles. With integration of electric mobility, one of the major future use cases and its special set of requirements is thoroughly discussed.

12.1 Introduction

Today's electric power distribution systems are on the move to smarter grid infrastructures, accompanied by a complete restructuring of the conventional roles in energy markets. The modern-day architecture of electric power grids defines distinct roles for energy producers, suppliers, and consumers. With the new paradigm of smart grids driving toward sustainability, some of these market roles are redefined.

The energy supplier systems have to handle an increasing amount of energy gained from renewable energy sources, like wind and photovoltaics, as well as household energy production systems, like combined heat and power. Compared to traditional forms of energy supplies, these forms of energy are fed into the grid in a decentralized way, and furthermore underlie much more volatile supply characteristics. However, the key requirements for efficient and economically feasible operation of today's power plants are balanced and nonvolatile load levels. Both aspects directly influence the distribution process of transport and distribution system operators and require the adoption of advanced *information and communication technologies (ICT)* in these processes.

The role of classical customers also changes from pure consumers to more complex customers called *prosumers*.[1] The prosumer has the ability to feed energy back into the grid with his own generators. By enabling smart metering at the prosumer's site and providing different tariff models with load level-aware price incentives, the prosumer is actively involved in the energy trading market and, most importantly, becomes more aware of his power consumption levels. In order to avoid this new market involvement being interpreted as an additional burden for the prosumer, autonomously acting agents for managing and remote controlling a smart home environment play a key role in this regard. For this purpose, the prosumer needs transparent and bidirectional connectivity to the regional energy marketplace. This comes along with additional requirements on *data rates, quality of service (QoS)*, and *security policies* with respect to the underlying ICT architecture. Figure 12.1 shows the resulting scenario and provides an overview on relevant system levels: *home area networks (HANs), neighborhood area networks (NANs), networks of distribution/transport system operators (DSO/TSO), regional energy marketplaces*, and *energy trading markets* like *European Energy Exchange (EEX)*.

This work focuses on the design of an ICT reference architecture for the outlined scenario. The progress from today's power grids toward ICT-driven smart grids, taking numerous distributed and renewable energy resources into account, is a long-term transition and will build upon continuous technological advancements in both power electronics and ICT sectors. Additionally, due to the short technology life cycles of ICTs, it is of particular importance that proposed architectures are based upon well-proven standards for interoperability as well as providing means for adaptability to new requirements and technology advancements. Hence, interoperability between a wide spectrum of communication technologies and protocols should be supported by using standardized, open communication technologies and Internet protocols. An overview on state of the art is given in Section 12.2.

The proposed reference architecture is introduced in Section 12.3. Special focus is placed on the regional energy marketplace that consolidates all transactional information sources. It is the central ICT anchor point for future extensions providing means for information provisioning of infrastructural enhancements as well as new actors offering value-added services to other participants of the marketplace. It furthermore provides a platform for new market participants with moderate market entry barriers, and therefore enables them to become part of the energy market value chain.

---

[1] **pro**ducer and con**sumer**

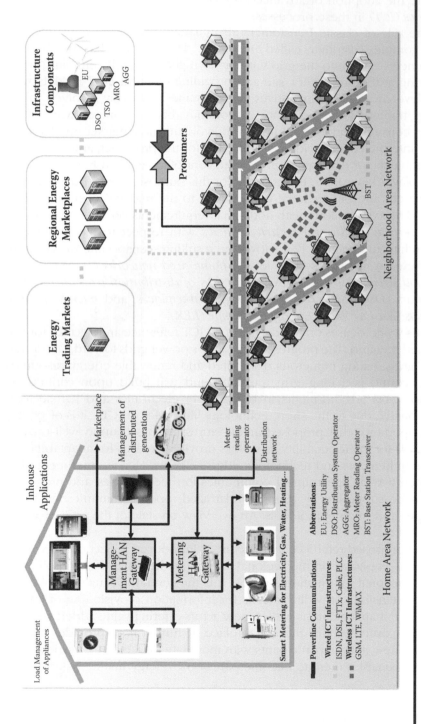

Figure 12.1 Overview of system levels in smart grid scenario.

The section also focuses on additional resources that are required for integration and consideration of customer appliances and assets via local ICT *HAN gateways*.

Electric vehicles (EVs) are considered one of the major drivers toward smart grids. However, charging EVs also necessitates some special requirements. These requirements and the integration of relevant stake holders into the proposed system architecture are discussed in Section 12.4. Two major use case scenarios are depicted: *private home charging* and *public charging*.

12.2 State-of-the-Art ICT for a Smart Grid

This section gives an overview on current state of the art for Internet and communication technologies and illustrates current trends in mapping these technologies to the smart grid. Related work on ICT architectures for power grids and common standards for grid communication systems are briefly discussed in Section 12.2.1. A more detailed comparison of applicable technologies/protocols categorized by different associated ISO/OSI layers is given in Sections 12.2.2, 12.2.3, and 12.2.4.

12.2.1 Systems and Infrastructures

Several approaches for interoperable communication system infrastructures for future energy grids are presented in literature [1–4]. Interoperability between different technologies and architectures regarding energy distribution and ICT is the driver for defining reference architectures resulting in abstract component descriptions and information exchange patterns [1]. An overview of the current status of standardization is given in Basso and DeBlasio [2]. The reference architecture presented in Beer et al., [3] and [4] describes an abstract system architecture being the basis for subsequent detailed system specification. The presented approach is based on international standards, like IEC 61968, [5], IEC 61970, [6] and IEC 61850, [7] offering reliable concepts for management and automation functionality. A reference architecture for object models, services, configuration languages, and protocols has been established by the IEC TC 57 [8]. It associates and consolidates different standardization efforts in order to define a *Seamless Integration Architecture*. One result of this work is the IEC 61850 standard providing a framework for energy automation. An adjacent standard—IEC 62351 [9]—targets dedicated security requirements and provides security solutions for this application domain.

The architecture presented in this chapter is derived by a technology-independent and requirement-driven approach. Therefore, an in-depth requirements analysis on the communication interfaces and their information exchange patterns was conducted. This strategy enables selection of suitable technologies for different interfaces and a reliable system design. Besides the exchange of automation and management information, tariff and measurement information is communicated. The latter have different needs in terms of reliability, latency, and data rate, but also regarding their security requirements, e.g., confidentiality or nonrepudiation.

12.2.2 Lower Layers

The physical transmission of the data in a smart grid environment is based upon the particular use case scenario with different quality of service requirements. Due to the large diversity of technology life cycles between the electric and ICT components, it appears that a compromise between the system capabilities and an economic point of view is stringently required. Especially the choice of an appropriate media layer affects the overall system design, and with that the efficiency and economy of the infrastructure. A comparison of established and future transmission technologies in the smart grid context for in-house and wide area communication technologies is given in Sections 12.2.2.1 and 12.2.2.2.2.

12.2.2.1 In-House Communication Technologies

By establishing an ICT infrastructure for metering gateways, enhanced energy management for decentralized power generation (e.g., solar panels, wind power, power-heat coupling) and Demand Side Management (DSM) (e.g., household appliances, electric vehicles [EVs]) are feasible for private households. The integration of these components combined with a decoupling of ICT and energy components comes along with the necessity for reliable in-house communication networks.

Several solutions are discussed for these use cases, like Wireless M-Bus [10], ZigBee [11], Z-Wave, [12] and KNX-RF [13] for transmission of metering data. For in-house energy management and home automation ZigBee [11], Wireless LAN [14] and narrowband RF systems are used. As wired technologies and bus systems, Broadband Powerline Communications (BPLC) [15], Ethernet [16], KNX [13], Konnex RF, [13] and M-Bus [17] are widely deployed.

In order to ensure transparent connectivity to all in-house components, the comprehensive introduction of HAN gateways is one of the central elements for combining the high demands on security and providing an extensive connectivity to the prosumer's household. On the one hand, the gateways provide firewall functionality; on the other hand, the gateways provide connectivity to all HAN entities. Thereby functionalities like smart metering for multiple metering devices, DSM for loads and decentralized power generation, as well as user interaction are provided by the gateway. The HAN gateway collects and stores metering data from several metering devices, such as electricity, gas, water, and heating meters using dedicated wireless technologies. The collected data are bundled and securely transmitted to the meter reading operator typically using wireless wide area point-to-multipoint technologies.

The energy management can be done either through the prosumer itself, motivated by tariff, or through the distribution network operator for controlled or emergency load reductions. Therefore, interfaces to the prosumer's appliances, loads, and local power generation components are provided by integrating these components into the prosumer's in-house network. Depending on external pricing information, several loads can be controlled, e.g., the charging process of EVs as well as controlling home heating systems. An intelligent washing machine makes use of the dynamic tariff information by starting the washing procedure in low-tariff periods and avoids starting it in high-level tariff periods. In addition to this, the connection interface can be used for maintenance, remote configuration issues, and firmware updates.

One of the key capabilities is the integration of decentralized power generation systems, which will make up a large part of the power generation in future systems. Nowadays an increase in local power generation installation can be observed, e.g., solar panels, wind power plants, and combined heat and power generation. Referring to the communication market, a dedicated infrastructure has to be provided by an operator that is also providing management and installation services to the prosumers. In order to maintain a reliable ICT infrastructure between the HAN gateways and the marketplace, a HAN gateway operator provides the reliable ICT components for software updates, administration, and configuration issues.

12.2.2.2 Wide Area Communication Technologies

In order to provide connectivity to the customer's premises equipment, several options have been introduced, which are mainly based on two scenarios: a dedicated access network infrastructure and a shared access network infrastructure. By using a shared network infrastructure, restrictions in terms of quality of service have to be accepted due to nonexclusive usage of the medium. On the other hand, this solution offers an economic alternative by using existing infrastructures.

Wired technologies in the smart grid context are used for integrating the prosumer's households and substations into the infrastructure. Since installation costs are higher than for wireless technologies, the integration of existing infrastructures is typically preferred.

One of the prevailing solutions for broadband Internet access has been established by Digital Subscriber Line (DSL), [18] which offers, besides high availability, a good pervasion and high data rates. DSL offers in the smart, grid context an economic solution for the last-mile access of the prosumers households, whereas the nonexclusive usage of the medium and interfaces to critical infrastructure components needs to be covered by reliable QoS and security mechanisms.

Compared to DSL, the scope of operation of optical networks (e.g., fiber-to-the-home/building/etc. (FTTx) [19], Gigabit Passive Optical Network (GPON)) is broadly defined by covering solutions for last-mile access as well as substation automation. Advantages of using an optical network in the smart grid context are given by long technology life cycles and electromagnet compatibility robustness as well as high ranges and data rates.

Using power lines as the transmission medium for the data has been discussed for decades and has been established successfully for the transport grid. Concerning present efforts on integrating new actors, like the prosumers, into a smart grid, Broadband PLC technologies (IEEE P1901, HomePlug 1.0/1.0 Turbo/AV/AV+, DS2, Panasonic HD) [15] become more relevant. Especially due to new capabilities like larger bandwidth, higher-modulation schemes, and notching filter, the BPLC technologies offer an economic solution for the communication infrastructure on several levels of the smart grid. Likewise, the narrowband PLC technologies [20] are discussed as a last-mile solution due to moderate installation costs and exclusive usage.

Wireless technologies in general offer a cost-efficient solution for new communication infrastructures due to the missing installation costs for cables. On the other side, wireless technologies are based on a valid network design covering resource and frequency allocation.

Cellular networks like Global System for Mobile Communications (GSM) [21] and Universal Mobile Telecommunications System (UMTS) [22] are widely deployed, which provides a suitable solution for services with limited requirements on data rate, latency, and availability. Several smart metering projects are based upon these approaches, but for a comprehensive installation of smart meters and for offering enhanced services like DSM, enhancements to these technologies are necessary. Therefore, extensions to these technologies have been introduced already by defining General Packet Radio Service (GPRS) [23], Enhanced Data Rates for GSM evolution (EDGE), [24] and High-Speed Down/Uplink Packet Access (HSxPA) [25].

The next generation of cellular networks has been introduced by technologies like Worldwide Interoperability for Microwave Access (WiMAX) [26], Long-Term Evolution (LTE) [27], and LTE–Advanced, [28] covering requirements on higher data rate and real time. In the context of smart grid solutions, these technologies promise to meet the requirements for integrating prosumers' households as well as substation automation. Especially the usage of lower-frequency ranges for dedicated services (smart metering, demand side management, substation automation) offers a promising solution, whereas the development of future network deployments needs to be taken into account.

Furthermore, several approaches investigate the integration of established radio technologies like satellite technologies (Inmarsat, Iridium, Globalstar, Thuraya, Astra) and terrestrial trunked radio (TETRA) [29]. Due to a good penetration of satellite technologies, these approaches are taken into account to offer connectivity to rural areas. Limited data rates and current tariffs restrain a comprehensive usage of these technologies for advanced smart grid functionality.

12.2.3 Intermediate Layers

Services based on open and standardized Internet protocols have revolutionized the information and communication technology and provided new business opportunities leading to an enormous raise in productivity in the field of telecommunication and multimedia. Today's energy management systems can benefit to a large extent by this development. Internet protocols for the smart grid are summarized by the Internet Engineering Task Force in *Internet Protocols for the Smart Grid* [30]. The draft provides an overview of the Internet Protocol Suite (IPS) and the key infrastructure protocols that are critical in integrating smart grid devices into an IP-based infrastructure. On the transport layer the usage of User Datagram Protocol (UDP) [31], Transmission Control Protocol (TCP) [32], Datagram Congestion Control Protocol (DCCP) [33], and Steam Control Transmission Protocol (SCTP) [34] is described in the context of smart grid operations. In order to provide essential data exchange on the network layer, Multiprotocol Label Switching (MPLS) (RFC 3031) and Asynchronous Transfer Mode (ATM) are discussed, and a general usage of IPv6 is strongly recommended. Security issues are covered by discussing the usability of Virtual Private Networks (VPNs) and IPsec.

12.2.4 Higher Layers

The basic idea behind Service-Oriented Architectures (SOAs) is to automate business processes by shifting them to pure machine-to-machine communication. SOA is not a concrete technology but a recommendation for software design in a succession of object-oriented and component-based programming. Web services are the most popular implementation of SOA principles and are currently penetrating the field of business-to-business (B2B) processes. This also leads to one of the main advantages of Web Services, which is interoperability even in heterogeneity software and hardware environments. It is primarily achieved by adopting open and standard technologies like Extensible Markup Language (XML) as a common

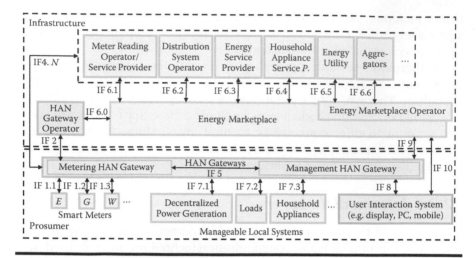

Figure 12.2 ICT reference architecture design.

denominator. Hence, the description of service interfaces or message interactions is interpretable by various endpoints and also human readable. Besides the advantages, such an approach also introduces huge overhead in terms of exchanging and processing of messages. These characteristics have been considered for specific application scenarios when embedded devices are involved in the network infrastructure (e.g., smart meters, home appliances, Wireless Sensor Networks (WSNs), etc.).

The Device Profile for Web Services (DPWS) [39] implements a subset of WS-* specifications in order to make the advantages of Web Services architectures available to the growing market of embedded systems. DPWS is the successor of the Universal Plug and Play (UPnP) [40] protocol suite. Both protocol stacks offer the same functionality to the user: addressing, discovery, description, control, eventing, and presentation of devices and their encapsulated services. The major advancement of DPWS over UPnP is the strict adoption of standard Web Services specifications. In contrast, UPnP defines very specific, not widely adopted protocols for device/service discovery and eventing. The strict adoption of WS-* specifications makes DPWS very attractive in the industrial domain because costs for integrating device processes into existing Web Services backbones are minimized. This is particularly useful in the industrial automation domain, where successful completion of business processes or their failure can be automatically incorporated and considered in the company's ERP, conveyed to other business processes or even customers.

12.3 ICT Reference Architecture Design

The reference architecture shown in Figure 12.2 adopts architectural concepts of related architectures and adds several enhancements to these approaches with respect to the related requirements described in Section 12.1. The purpose of the abstract description of the architecture is to identify relevant information flows and respective interfaces, e.g., between the HAN gateways and energy marketplaces, in order to specify appropriate ICT.

The presented architecture is driven by two major components: regional energy marketplaces and local dedicated HAN gateways in each household.

12.3.1 Regional Energy Marketplaces

The central element of future energy systems is based on regional energy markets enabling the prosumers to manage their contracts, use advanced services of third-party service providers, and trade their energy consumption and generation contracts. Each energy marketplace is operated and maintained by a marketplace operator. Due to the interdependencies (e.g., in terms of balancing power) of multiple regional marketplaces, each managed by its own regional operator, interoperability between these platforms is one of the core requirements of the marketplace design. Therefore, these interfaces are designed to support typical standards of the application domain, which often stem from ISO/IEC. These standards commonly refer to well-established Internet standards like XML.

12.3.2 Local HAN ICT Gateways

In order to ensure transparent connectivity to all in-house components, the comprehensive introduction of HAN gateways is one of the central elements for combining the high demands on security and providing an extensive connectivity to the prosumer's household. On the one hand, the gateways act as firewalls, and on the other hand, the gateways provide connectivity to the HAN entities. Thereby two different forms of HAN gateway can be considered:

- Metering HAN gateways
- Management HAN gateways

The *metering HAN gateways* collect and store metering data from several metering devices, like electricity, gas, water, and heating meters. The collected data are securely transmitted bundled to the meter reading operator. Security in this context comprises two services: integrity protection and nonrepudiation for metering data, which is often done by using digital signatures allowing the meter reading operator to validate that the metering data have not been altered by an intermediate component. Furthermore, the bundled transfer may be encrypted to ensure the privacy of user related data in environments, where physical access to the transmission path cannot be secured. Moreover, a local feedback system may give the prosumer transparent insight into his current energy consumption. In conjunction with available tariff information, motivation for reducing overall power consumption can be achieved.

The *management HAN gateway*, which can be realized as an integrated HAN gateway or as a separate hardware component, represents an enhancement of the metering HAN gateway and offers management functionality to the prosumer's HAN entities. The energy management can be done either through the prosumer itself, motivated by tariff or sustainability, or through the Distribution System Operator (DSO) for controlled or emergency load reductions. Therefore, interfaces to the prosumers' appliances, loads, and local power generation components are provided. Depending on an external pricing information, several loads can be controlled, e.g., charging of EVs or controlling a home heating system. For example, an intelligent washing machine makes use of the dynamic tariff information by starting the washing procedure in low-tariff periods and avoids starting it in high-level tariff periods. In addition to this, the connection interface can be used for maintenance, remote configuration issues, and firmware updates.

The reference architecture presented in the previous section introduced the integration of both HAN gateways at the prosumer's premises. In general, three modes of operation are presented in this section that combine various components of the reference architecture differently:

■ Mode 1: Joint HAN gateway operator
■ Mode 2: Separated HAN gateway operator
■ Mode 3: Split HAN gateway

In modes 1 and 2 only one *HAN gateway* exists. Mode 1 provides the *HAN gateway operator* and the *meter reading operator* at the same instance, which ensures the operational infrastructure between the HAN gateway and the remaining service infrastructure. Compared to this, mode 2 explicitly separates the *HAN gateway operator* and *meter reading operator*. Hereby an independent *HAN gateway operator* is responsible for the operational communication infrastructure between the consumers' gateways and the remaining service infrastructure. The gateway gets installed, configured, and administrated by the *HAN gateway operator*. The acquisition of measurement data is done through a dedicated infrastructure, which is provided by the *HAN gateway operator*.

The *HAN gateway* in mode 3 is split into two physical entities, the *metering HAN gateway* maintained by the *HAN gateway operator* and the *management HAN gateway* maintained by the customer himself. The metering gateway is installed, configured, and administrated by the *HAN gateway operator*. By using this operational scenario, both gateways are separated and only linked by interface IF 5, which is used for administrative purposes in the management HAN gateway and for transmission of power consumption data to the user interaction system. The gateways may be virtually separated or physically separated, depending on the underlying deployment scenario.

12.3.3 Marketplace Participants and Interfaces

The infrastructure components consist of the following B2B systems interacting with the prosumers' equipment and other components via the marketplace platform:

- Meter reading/using operator (MRO/MUO)
- Distribution system operator (DSO)
- Transmission system operator (TSO)
- Energy service provider (ESP)
- Electric utility (EU)
- Appliance service provider (ASP)
- Aggregator (AGG)

In order to limit the number of interfaces between these participants, multiple interfaces between groups of components are combined together; e.g., interfaces between meter and HAN gateway are combined with one single interface IF 1.N, and interfaces between local systems and the HAN gateway are labeled IF 7.N. The system specification builds upon these groups of interfaces (Table 12.1) and specifies the necessary subsets of the interfaces for each component individually. Multiple access technologies are possible in order to ensure communication between infrastructures on the prosumer's and backbone's side. On the one hand, dedicated infrastructures ensure high quality of service due to an exclusive use (e.g., broadband PLC infrastructure [41]). On the other hand, shared public access technologies (e.g., GPRS or DSL) offer more cost-efficient solutions.

12.4 Integrating Electric Mobility in Marketplace Infrastructures

This section details the integration of electric mobility into the proposed architecture for regional energy marketplaces. In order to provide a common basis for the integration approach, the requirements are described in the following section. This results in the introduction of new stakeholders in

Table 12.1   Ten System Interfaces/Groups of Interfaces

Interface	Description
IF 1.N	Metering interfaces
IF 2.N	Administrative interfaces
IF 3.N	Marketplace interfaces (future usage)
IF 4.N	Service interfaces (HAN gateways)
IF 5.N	Inter-HAN gateway interfaces
IF 6.N	Service interfaces (e-energy marketplace)
IF 7.N	Interfaces between HAN gateways and manageable prosumer loads
IF 8.N	User interfaces (HAN gateway)
IF 9.N	HAN gateway interface for the e-energy marketplace
IF 10.N	User interfaces (e-energy marketplace)

Section 12.4.2. The corresponding adaptation of the previously proposed architecture for integrating electric mobility is described against the background of predominant use cases for private home charging and public charging in Section 12.4.3. Especially with respect to public charging, the mobility management of EVs plays a major role, which is discussed in Section 12.4.4.

12.4.1 Use Cases and ICT Requirements for Electric Mobility

The development of EVs is one of the major core drivers for establishing smart grids. EVs being connected to the grid provide energy storage capabilities that do not exist in today's distribution network levels. Moreover, looking at the number of hours a vehicle is typically parking throughout a whole day, EVs would eventually provide solid means for DSM. Both aspects are core enablers for introduction of renewables into power grid infrastructures, but also ultimately lead toward additional need for ICT-based coordination in power distribution networks.

With regards to this hypothesis, electric mobility has some special requirements compared to traditional electric consumers that must be considered for ICT integration tasks. Two major use cases are differentiated in nowadays discussions on electric mobility: *private home charging* and *public charging* of EVs.

Looking at private home charging, EVs are not permanently connected to the household's infrastructure, and therefore have different availability patterns than other appliances, like refrigerators or air conditioning. However, the majority of journeys with road vehicles follow regular patterns. Hence, charging an EV could be parameterized with these patterns. Nevertheless, the system should be able to provide some reserve capacities for spontaneous use. This already indicates that the underlying charging system must be designed in a user-centric way and must be integrated as seamless as possible in order to provide acceptable means for replacing today's combustion engine-based road vehicles for individual mobility. Hence, integration of *electric vehicle supply equipment (EVSE)* into previously introduced HAN gateway environments is a core requirement for harmonizing technical requirements with individual needs for private home charging. The integration with proposed HAN gateways furthermore enables DSM-related use of EVs coordinated by verified and approved stakeholders on the energy marketplace and beyond the vehicle owner's use.

In case of public charging, further requirements must be considered: due to the mobility of EVs, a customer's overall energy demand cannot be determined solely by stationary meter readings at the customer's site. Especially on longer-distance runs, the power for charging an EV is consumed along the owner's mobility path. Hence, for a contract-based customer relationship, the meter readings together with some identification credentials must be communicated from public EVSEs to respective accounting entities. These identification credentials include both accounting information and authorization for gaining access to charging infrastructures. The latter clearly indicates the imminent need for open and interoperable charging infrastructures with roaming capabilities between various infrastructure providers. Patronizing users in this regard would lead to major acceptance issues for electric mobility.

12.4.2 Electric Mobility Roles and System Model

In order to align to the previously identified core requirements for private home charging and seamless public charging, this section introduces the currently discussed role model for electric mobility. Relevant roles and components as well as main information flows between these are defined in this section and presented in Figure 12.3.

Three distinct domains are identified in the context of this work: *user/customer*, *electric mobility*, and *utilities*. Each domain indicates its assets by incorporating specific roles or components. Altogether, they constitute the investigated role-based system model for electric mobility.

The *customer* domain incorporates all customer-related assets. These are the *user*, the *electric vehicle (EV)*, and any type of *information client(s)*. Figure 12.3 furthermore depicts the charge spot, also called *electric vehicle*

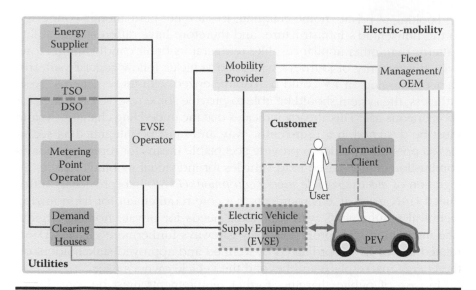

Figure 12.3 Role-based system model for electric mobility.

supply equipment (EVSE), as part of both the *customer's* and the *electric mobility's* domain, indicating various ownership options for EVSEs: private, semipublic, or public. The EVSE marks an endpoint of the electric grid. In case of conductive charging, the EV is connected to an EVSE with a charge cord and communicates through the vehicle-to-grid (V2G) interface (indicated with a red arrow in Figure 12.3). A fixed metering point for collecting electric meter readings is dedicated to each EVSE.

The *electric mobility* domain incorporates the *EVSE operators,* who provide the charging infrastructure to the *customer* and are responsible for the operation and maintenance of the EVSEs. The *EVSE operator* is the power recipient for every single EVSE from *energy suppliers.* Furthermore, the *EVSE operator* takes the operational responsibility for providing access to its charging infrastructure. The *mobility provider* closes a contract with the *customer* (e.g., private customers or fleet customers) for electric mobility services. The provided services and also the means for accounting depend on the business model of the *mobility provider.* In case of the envisioned open markets, such a contract enables a customer to gain access to *any* EVSE operator's charging infrastructure. However, this assumes that each *mobility provider* must have means available to negotiate with each and every *EVSE operator* in order to enable seamless roaming for his customers. In the following section we will focus on this issue, taking into account the energy marketplace integration. The *fleet management* role extends the *mobility provider* role and acts upon battery management information from groups of EVs (e.g., targeted loading profile, location and time, etc.), as well as the forecast of estimated energy demand for groups of EVs.

Such a role would enable balancing power based on the consolidation of demand/supply capacities of large fleets of EVs. The *fleet management* is not mandatory for controlled flow of energy but could provide an added value for optimized energy efficiency. The *electric mobility* domain might furthermore contain other value-added service providers.

The *utilities* domain includes hierarchically organized *demand clearing houses (DCHs)*. They provide means for autonomous load leveling on different grid levels (e.g., substation level, DSO, TSO). Further roles in this domain are taken from traditional grid infrastructures, like *metering point operators (MPOs)*, *distribution system operators (DSOs)*, *transport system operators (TSOs)*, and *energy suppliers*, which were already introduced in Section 12.3.3.

12.4.3 Electric Mobility Roles in Energy Marketplaces

The regional energy marketplace architecture described in Section 12.3.3 must be extended by those roles defined in the previous section. Figure 12.4 illustrates how seamless integration could be established in line with previously identified requirements from Section 12.4.1.

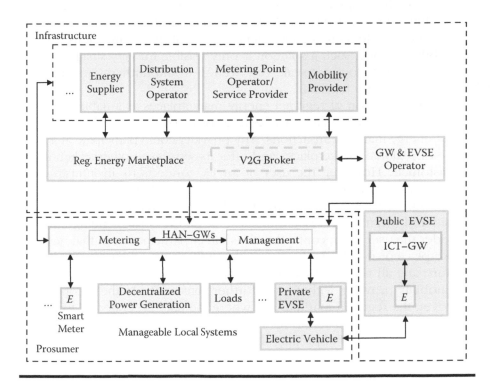

Figure 12.4   Electric mobility extensions.

In case of private charging, the EVSE becomes part of the prosumer's HAN and communication to the marketplace is not directly established by the EVSE, but through the *HAN gateway*. This allows us to keep complexity of private EVSEs at a minimum and reduces costly wide area link redundancies. In both use cases the marketplace integration furthermore supports decoupling of mobility providers and customers supporting competitive market environments. However, establishing regional marketplaces requires additional management regarding mobility aspects in case of public charging. Users might roam between various regional marketplaces, which needs to be considered.

12.4.4 Electric Vehicle Mobility Management

For contract-based charging, the mobility of EVs, their spatially unpredictable demand, and the heterogeneity of infrastructures and providers require additional mechanisms in order to ensure interoperability.

A potential solution for interoperability between various EVSE operators and mobility providers would be to close infrastructure access agreements. Legal contracts as well as network charge contracts and a common agreement on B2B interfaces would be necessary in order to fulfill such an agreement. The problem with this approach is that with increasing numbers of mobility providers (M) and EVSE operators (N), the necessary amount of agreements would increase according to $M * N$ if we assume holistic interoperability. Hence, such a solution is only feasible for very small numbers of providers and does not scale very well. It would furthermore introduce major market entry barriers.

The introduction of regional energy marketplaces as a liable entity introduces an abstraction layer between various providers, and therefore simplifies energy market transactions. Any provider could participate at a multitude of regional marketplaces and offer its services to other stakeholders or customers. However, the user of an EV connects to an infrastructure that is always assigned to exactly one regional marketplace. According to this primitive, there are basically two different scenarios that need to be considered when an EV user wants to charge: the corresponding mobility provider the user has a contract with is either participating in that regional marketplace or not. In the first case, the authorization to access the infrastructure and all other service transactions can be directly negotiated with the mobility provider without any further constraints. In the second case, however, the regional marketplace where the user is connected to, needs to negotiate a liable external relationship. Here again, two options apply: Similar to the original approach, the mobility provider establishes roaming agreements with other mobility providers. In that case, the roaming partner delivers the service to EV users even if its original contractor does not participate in the same market. The second option involves inter-market transactions that depend on less numbers of entities as opposed to the individual market participants. The management of intermarket transactions for EV charging is inspired by the roaming approach already known from mobile communication infrastructures. Roaming for users of EVs is provided by two new *virtual* entities:

- Home location marketplace (HLM)
- Visited location marketplace (VLM)

Each entity is registered in the V2G broker of each regional energy marketplace. The HLM represents the regional energy marketplace where the user is originally registered. In most cases this marketplace should be equivalent to the market to which the household of the customer is registered. The VLM represents the energy marketplace the customer is joining by plugging in his EV in the infrastructural domain of a marketplace different from his HLM and without any roaming partner. During the authentication process the EV communicates to its corresponding HLM through the vehicle-to-grid communication interface to the VLM. The VLM now requests synchronously all relevant information from the HLM in order to provide access to the power grid for the customer. After the charging process the VLM reports all necessary accounting information to the MSP residing at the HLM asynchronously. Similar to mobile communications the VLM accounts for roaming fees that correspond to network charges and which are necessary in order to provide liable means for balancing grid demands.

In contrast to the mobile communication architecture, it is expected that from a security point of view the underlying architecture will make use of certificates and digital signatures to ensure integrity and nonrepudiation for energy offers and concluded charging contracts, which serve as the base for accounting and billing.

12.5 Conclusion and Outlook

Based upon a thorough state-of-the-art review of information and communication technologies in the context of smart grid, a reference architecture was presented and discussed. The proposed reference architecture design represents the architectural concepts for high-level interaction between particular entities of the ICT infrastructure of future energy grids. The requirements and definition of interfaces toward a regional energy marketplace have been discussed by a technology-independent approach. Extension points as well as adaptability of the presented concept have been presented by discussing the integration of electric mobility components. Furthermore, the trends toward mobility management for electric vehicles based on existing mobile communication roaming approaches was elaborated. Such mobility management systems would allow TV users to roam between distinct charge point operators in highly scalable deployment scenarios.

Acknowledgment

The work in this chapter was partially funded by the German Federal Ministry of Economics and Technology (BMWi) through the project *E-DeMa* with reference number 01ME08019A. The authors thank the project partners RWE, Miele, ProSyst, SWK, and ef.Ruhr for the discussions within the project. The content of Section 12.4 demonstrates the extensibility of the proposed architecture and does not reflect the opinion of the consortium.

References

1. IEEE P2030/D2.1 Draft, *Draft guide for smart grid interoperability of energy technology and information technology operation with the electric power system (EPS), and end-use applications and loads.* (IEEE, New York, May 2010).
2. T. Basso and R. DeBlasio, Advancing smart grid interoperability and implementing NISTs interoperability roadmap: IEEE P2030TM initiative and IEEE 1547TM interconnection standards, in *Grid-Interop 2009 Conference* (November 2009).
3. S. Beer, H. Rüttinger, L. Bischofs, and H.-J. Appelrath, Towards a reference architecture for regional electricity markets (entwurf einer referenzarchitektur für regionale elektrizitätsmärkte), *Information Technology*, **52**(2), 58–64 (2010).
4. S. Beer, L. Bischofs, C. Pries, M. Uslar, A. Niee, H.-J. Appelrath, M. Rohr, and M. Stadler, The eTelligence reference architecture—A standard-based architecture for regional electricity markets, in *Internationaler ETG-Kongress 2009*, VDE Verlag (October 2009), pp. 84–89.
5. IEC 61968, *Application integration at electric utilities—System interfaces for distribution management—Part 11: Common information model (CIM) extensions for distribution ed. 1.0* (IEC, Geneva, Switzerland, July 2010).
6. IEC 61970, *Energy management system application program interface (EMS-API)—Part 301: Common information model (CIM) ed. 2.0* (IEC, Geneva, Switzerland, April 2009).
7. IEC 61850, *Communication networks and systems for power utility automation—Part 7-4: Basic communication structure—Compatible logical node classes and data object classes ed. 2.0* (March 2010).
8. IEC TC57/TR62357, *Power system control and associated communications: Reference architecture for object models, services and protocols* (IEC, Geneva, Switzerland, July 2003).
9. IEC 62351, *Power system control and associated communications—Data and communication security—Part 1: Communication network and system security: Introduction to security issues* (IEC, Geneva, Switzerland, May 2007).

10. *DIN EN 13757-4:2005, Communication systems for meters and remote reading of meters—Part 4: Wireless meter readout (Radio meter reading for operation in the 868 MHz to 870 MHz SRD band)*, (DIN EN, 2005).
11. IEEE 802.15.4-2006, *ZigBee*.
12. Zensys, Z-WAVE Alliance, *Z-WAVE*.
13. *ISO/IEC 14543-3, Information technology—Home electronic system (HES) architecture* (2010).
14. IEEE 802.11/b/b+/11g/11n, *WLAN*.
15. IEEE P1901, *BPLC—Broadband powerline communication*.
16. *IEEE 802.3, Ethernet working group, http://www.ieee802.org/3/*.
17. EN 13757-4, *Communication systems for meters and remote reading of meters—Part 4: Wireless meter readout (Radio meter reading for operation in the 868 MHz to 870 MHz SRD band (MBUS)*.
18. ITU-T G992.1 (ADSL Spezifikation), ETSI TS 101952-1-4 (Spezifikation fr ADSL ber ISDN oder POTS), ANSI T1.413-1998 (Erste standardisierte ADSL Spezifikation), ITU-T G992.3 (ADSL2), ITU-T G992.5 (ADSL2+), *xDSL*.
19. EN 188 000, Fachgrundspezifikation: Lichtwellenleiter, *FTTx—Fibre to the X* (ITU—International Telecommunication Union).
20. IEC 61334-x-Steuerungs-und Messaufgaben, *NPLC—Narrowband powerline communication*.
21. 3GPP—3rd Generation Partnership Projekt, *GSM* (3GPP, TS 41.001—GSM specification set).
22. 3GPP—3rd Generation Partnership Projekt, *UMTS* (3GPP, TS 25.101).
23. 3GPP—3rd Generation Partnership Projekt, *GPRS* (3GPP, TS 45.001—physical layer on the radio path General description).
24. 3GPP—3rd Generation Partnership Projekt, *EDGE* (3GPP, TS 43.051—GSM/EDGE radio access network (GERAN) overall description; stage 2).
25. 3GPP—3rd Generation Partnership Projekt, *HSxPA* (IEEE, TS 25.308—high speed downlink packet access (HSDPA); overall description; stage 2; TR 25.999—high speed packet access (HSPA) evolution; frequency division duplex (FDD)).
26. IEEE 802.16, *WiMAX*.
27. 3GPP—3rd Generation Partnership Projekt, *LTE—Long term evolution*. (3GPP, TS 36.101—evolved universal terrestrial radio access (E-UTRA); user equipment (UE) radio transmission and reception).
28. 3GPP—3rd Generation Partnership Projekt, *LTE—Advanced* (3GPP, TR 36.806—evolved universal terrestrial radio access (E-UTRA); relay architectures for E-UTRA (LTE-advanced)).
29. ETSI EN 300 392-2 v3.2.1, *Tetra*.
30. F. Baker, D. Meyer, *Draft-core-13: Internet protocols for the smart grid* (IETF, March 2011).
31. J. Postel, User datagram protocol, RFC 768 (standard) (August 1980), http://www.ietf.org/rfc/rfc768.txt.
32. J. Postel, Transmission control protocol, RFC 793 (standard) (September 1981), http://www.ietf.org/rfc/rfc793.txt. Updated by RFCs 1122, 3168, 6093.
33. E. Kohler, M. Handley, and S. Floyd, Datagram congestion control protocol (DCCP), RFC 4340 (proposed standard) (March 2006), http://www.ietf.org/rfc/rfc4340.txt. Updated by RFCs 5595, 5596.

34. R. Stewart, Stream control transmission protocol, RFC 4960 (proposed standard) (September 2007), http://www.ietf.org/rfc/rfc4960.txt. Updated by RFC 6096.

35. B. Adamson, C. Bormann, M. Handley, and J. Macker, NACK-oriented reliable multicast (NORM) transport protocol, RFC 5740 (proposed standard) (November 2009), http://www.ietf.org/rfc/rfc5740.txt.

36. IEC 62056-42, *Physical layer services and procedures for connection-oriented asynchronous data exchange, DLMS.*

37. IEC 62056-42, *Physical layer services and procedures for connection-oriented asynchronous data exchange, Smart message language,* (November 2008).

38. *ZigBee Smart Energy Profile 2.0.*

39. A. Regnier, D. Driscoll, and A. Mensch, *Devices profile for web services, version 1.1* (OASIS, http://zigbee.org, July 2009).

40. A. P. et al., *UPnP device architecture 1.1.* (UPnP Forum, 2008).

41. IEEE P1901 draft, *Draft standard for broadband over power line networks: MAC and PHY specifications* (IEEE, New York, January 2010).

Index